Teaching metric awareness

The decimeter, liter,
kilometer, and gram
Will cause a lot of problems
For our dear old Uncle Sam
But oh how pleasant to perceive
that metric goals and aims,
Can be furthered and abetted
by activities and games.

M. J. Kurtz

Teaching metric awareness

V. Ray Kurtz

Department of Curriculum and Instruction
Kansas State University
Manhattan, Kansas

With 68 illustrations by Michael Carpenter

The C. V. Mosby Company

Saint Louis 1976

Printed in the United States of America

Distributed in Great Britain by Henry Kimpton, London

Library of Congress Cataloging in Publication Data

Kurtz, Vernon Ray, 1933-
Teaching metric awareness.

SUMMARY: Introduces the principles of the metric system and their application in everyday situations.
1. Metric system—Study and teaching—United States. 2. Weights and measures—United States. [1. Metric system. 2. Weights and measures.] I. Title.
QC93.K87 389'.152 75-22174
ISBN 0-8016-2811-3

GW/M/M 9 8 7 6 5 4 3 2 1

WHO CAN USE THIS BOOK

If serious thought is given to the changes that will be necessary with the coming adoption of the metric system and the accompanying difficulties, it becomes quite evident that few people in this country are really prepared to use the metric system. If 100 people were stopped on the sidewalk and asked their height in centimeters, the variability in answers would be humorous. If each were asked, "What is your weight in kilograms?" the answers might range from 10 to 1,000. The few exceptions to the previous examples would be those persons who, through considerable experience in science, had gained the skill of converting rapidly from the traditional system to the metric system.

Over a decade ago, "new mathematics" made vast differences in the teaching of mathematics. However, the brunt of the revolution was felt by teachers who had to learn the new methods and by frustrated parents of school children who tried to assist with homework. The incursion of the metric system will be distinctly different in scope. No one will be exempt from participating in this revolution. One who does not participate may be compared to the person who crosses a time zone and will not change his watch. He will be handicapped in interacting with those around him. Thinking metric will possibly be more troublesome for mature learners than for children, since adults are deeply entrenched in the English measurement system.

In consideration, then, of the fact that it will be so difficult for most people to think metrically, this book is designed to build an adequate competency in the metric system by providing activities that will progress from concrete to semiconcrete to abstract. The book is designed to provide those strategies that may be used by the mature learner as well as by children to provide a feeling of ease with the new system. The organization is aimed at providing those procedures that not only will be helpful to teachers preparing to teach the metric system but will also be appropriate in teaching children. Despite the seeming diversity of these two groups, the strategies presented in this book can be adapted to fit both because it will be necessary for both groups to start at the beginning. Regardless of age, a learner must begin with such concrete activities as measuring, weighing, and pouring. Adults will bring a greater maturity to the experience than will children; however, younger learners will be free from the inhibitions and prejudices that may hinder the adults.

Thus, this book is written for learners of any age who wish to master the metric system of measurement. If those who desire to teach the metric system will proceed through the chapters, participating in the activities as well as preparing materials for use with children, the end result will be metric awareness for the teacher as well as activities to use with children.

WHAT THIS BOOK CONTAINS

After "Getting Started" (which includes a brief review of the conditions that are precipitating the worldwide changeover from the English to the metric system and the rationale for the change), the metric components of length, weight, area-volume, time, and temperature are treated in separate chapters. The pre-service and in-service teacher will find at the beginning of each chapter a discussion that presents the rationale for the strategy to be used in teaching the specific metric topic. Woven into each chapter are the various activities designed to provide those experiences necessary to develop metric awareness for each student.

The final chapter, entitled "Leftovers," contains activities that did not fit in the preceding chapters, bulletin-board ideas, and metric recipes. An appendix contains commonly used metric and English relationships and sources of metric materials. Virtually all of the activities presented in this book may be inexpensively and easily prepared. Many can be made by children. No classroom should be deprived of metric aids because of lack of funds. Although there are many expensive commercially prepared metric aids, a number of them have been hurriedly put on the market and thus may or may not perform the function intended by the classroom teacher.

A careful examination of the book will show that no activities involve conversion of units from metric to English or from English to metric. This is based on the firm belief that students of any age need physical experiences that familiarize them with the units of the metric system. I see no reason to provide conversion problems, since this makes the learner dependent on traveling through the English system in order to get to the metric system. However, as students become more sophisticated in their ability to work with metric units, they should be encouraged gradually to work with more abstract activities that follow the concrete activities presented in the following chapters.

In summary, I have maintained the following three priorities while writing this book:

1. Since students learning the metric system do not need a conversion approach, no conversion activities have been presented.
2. Since students learning the metric system need many experiences such as measuring, weighing, and pouring, the stress in the initial stages has been placed on specific activities.
3. Since students and teachers prefer that mathematics be enjoyable, the activities presented in this book are game-like in format.

CONTENTS

Appendixes

CHAPTER 1

GETTING STARTED

In 1971 the secretary of commerce of the United States transmitted to Congress the results of a 3-year study of the advantages and disadvantages of increased use of the metric system in the United States.[1] The report recommended that the United States change to predominant use of the metric system through a coordinated national program. This call was nothing new, as Thomas Jefferson first proposed that the United States go metric in 1790. However, this time the change appears imminent.

Most Americans are asking the question, "Why should the United States go to all the trouble and tremendous expense to change to a different system?" The answers to this question are legion. Possibly, the greatest advantage of the metric system is that, like the American monetary system, the units bear a decimal relationship to one another, which permits changing from one unit to another by moving the decimal point to the right or left depending on whether multiplication or division is involved. Consider the measurement of length where 342 centimeters is equal to 34.2 decimeters or 3.42 meters or 0.342 dekameters. This example is typical of the ease with which dividing by 10 may be accomplished in the decimally based metric system. By contrast, the English system of measurement, having evolved from various measurements corresponding to parts of the human body or other physical objects, not only uses nondecimally related units, but also uses a different one for each unit of measurement (for example, 12 inches to a foot, 3 feet to a yard, 5,280 feet in a mile, 16 ounces to a pound, 2,000 pounds to a ton). The ambiguity and confusion of the English system can be illustrated by the latter unit, as there are several tons. In addition to the regular ton (2,000 pounds), also referred to as the short ton, there is the long ton (2,240 pounds), and the freight ton, which varies according to the commodity being weighed.

An additional reason for change is that the United States is finding it difficult to stand alone in the world as the only major nation maintaining the English system. Even England, the nation that developed the "imperial" system of weights and measures, an-

[1] National Bureau of Standards: Brief history of measurement systems (Special Publication 304 A), Washington, D.C., 1972, Government Printing Office, p. 4.

nounced in 1968 that it would convert to the metric system. The timetable called for 1975 as the year for England to become a metric nation. The primary schools of England were required to adopt the metric system beginning in September 1969. In the United States, the pressure really began to be felt in 1973, when Canada and Australia announced that they too would begin the change from the English to the metric system. This left only the United States and a handful of smaller nations continuing with the English system. A further reason for metrication is the worldwide acceptance and use of the metric system in trade, for international trade standards, which are based on the metric system, act as a formidable barrier to any country that still operates on the English system. It is estimated that as much as $10 billion a year in additional trade can be expected after complete metrication.[2]

The greater precision and accuracy of the metric system is also considered to be an additional reason for adoption. Science and medicine have used this system for years, because the metric system had been adopted by the U.S. Health Department in 1902.[3]

STANDARDIZATION OF UNITS

One disconcerting factor in past years has been a lack of standardization of the metric system as to spellings and symbols for units. This will not clear up overnight, but there is a very promising international effort to standardize the metric units throughout the world. This system, *le Système international d'unites* (SI), is a modernized version of the metric system. It was established by international agreement to provide a logical interconnected framework for all measurements in science, industry, and commerce.[4] It was during the Eleventh General Conference of Weights and Measures (CGPM) that the name "international system of units" was adopted. The purpose of this group was to make recommendations on the establishment of a practical system of units of measurement suitable for international adoption. The SI base units as adopted by the CGPM in 1964, 1968, and 1971 are as follows[5]:

Quantity	*Name*	*Symbol*[6]
Length	Meter[7]	m
Mass	Kilogram	kg
Time	Second	s
Electric current	Ampere	A
Temperature	Kelvin	K[8]
Luminous intensity	Candela	cd
Amount of substance	Mole	mol

[2] "Soon it may be 'Give a centimeter and take a kilometer.' " Nation's Business **59:**89, April 1, 1971.

[3] Lottie Viets: Experiences for metric missionaries, The Arithmetic Teacher **20:**269, April 1973.

[4] National Bureau of Standards: op. cit., p. 2.

[5] National Bureau of Standards: The international system of units (SI) (Special Publication 330), Washington, D.C., 1971, Government Printing Office, p. 6.

[6] The symbols are given in lower-case letters unless the units are derived from proper names. Capital letters are used for the first letter of the symbol if it is derived from a person's name. The symbols are not followed by a period.

[7] There is no clear trend as to which spelling should be used, "metre" or "meter." In the United States the more common spelling has been "meter," whereas internationally the preferred spelling has been "metre." The teacher should emphasize to children that both spellings are correct.

[8] The Celsius scale and the Kelvin scale have the same-sized units. Since the Celsius scale is commonly associated with the metric system, it is included in this book.

All SI base units as adopted by CGPM were included in the preceding chart even though not all are developed in this book. The elementary teacher should be aware of the total list even though luminous intensity and amount of substance are normally beyond the scope of the elementary school. A teacher who is aware of these quantities will be prepared to assist a child should such a need arise. Derived units such as the square meter and cubic meter are included in this book as well as commonly used units such as the liter.

The CGPM has also adopted a standard list of SI prefixes, which, we hope, will be used on an international level. A partial listing of those most used are as follows[9]:

SI PREFIX

Prefix	*Factor*	*Symbol*
micro-	10^{-6}	μ
milli-	10^{-3}	m
centi-	10^{-2}	c
deci-	10^{-1}	d
deka-	10^{1}	da[10]
hecto-	10^{2}	h
kilo-	10^{3}	k
mega-	10^{6}	M

WHEN SHOULD METRICATION BEGIN?

Metrication should be started in the primary grades before the child has any opportunity to become biased against such a system. Teachers are finding that first-grade children enjoy properly organized metric activities. The main emphasis in the primary grades should be placed on concrete experiences that provide many actual measuring opportunities. In addition to concrete experiences, many recognition activities that afford the child an opportunity to recognize the printed words and symbols for the metric units, should be provided. Not all educators agree on whether primary school age children should be taught both the English and metric systems. Since it appears that both systems will be in use for the next several years, it appears to me that the systems should be taught simultaneously.

METRICATION FOR INTERMEDIATE AND OLDER CHILDREN

Even though intermediate and older children have had many experiences with the English system of measurement, they too are beginners with the metric measures. Therefore they must have beginning concrete activities. Similar strategies for teaching the metric system should be used for beginners regardless of age. Those students who have a basic understanding of the inch-pound system may be more resistant to a complete changeover. Teachers may find that presenting the advantages for the change may lessen the task. Even though children should be encouraged to think metric, they should not be chastised for wanting to refer to the meter as being a little longer than a yard. However, it is more desirable for them to think of a meter as 100 centimeters. The goal of metrication can be reached in a more natural manner if teachers will correlate the metric material with sci-

---

[9] National Bureau of Standards: The International system of units (SI) (Special Publication 330) Washington, D.C., 1971, Government Printing Office, p. 12.

[10] Note that the prefix for deka- appears as *da* in the SI symbols but *dk* in some metric literature. The *da* symbol is used throughout this book.

ence, social studies, sports, arts, cooking, and vocational work. All these areas abound with many metric opportunities.

GOAL OF METRICATION

As the metric change takes place, children are going to need ''bimensural'' skills involving both the metric and English systems. In comparison, a person who is truly ''bilingual'' is able to function with native-like control in two languages. He is comfortable and at ease with either language, slipping into one language with no reference to the other. This is quite different from the person who has only a reading knowledge based on translating skill. Those who have studied for a foreign-language reading examination are often unable to comfortably work with the second language. This translating skill is mainly helpful in passive usage.

Therefore, when the objective for the children of a nation is to become bimensural with the English and metric systems, native-like, active control becomes the goal. A person is not bimensural if he finds his height in centimeters by multiplying his height by 2.54. This skill would more accurately be classified as a translating skill. Native-like control enables the word ''metric'' to trigger thinking in metric terms with no reference to the other system. This can only be developed by working through a carefully designed sequence of learning activities or as a consequence of living with the new system.

CHAPTER 2

TEACHING METRIC LENGTH

The measurement of length is greatly simplified when the basic metric unit, the meter, and its multiples and submultiples are used. This is a contrast to the English system, which uses many unrelated units. Historically, the English system was based primarily on various dimensions of the human body. The inch is reported to have been originally defined as the length of the end joint of the adult thumb. As this became unworkable because of the great variation in adult thumbs, the inch was redefined as the length of three barleycorns taken from the middle of an ear of barley and laid end to end. Originally a foot was measured as the length of a man's foot. However this was later changed to conform to the barleycorn standard by a decree that fixed the foot at 36 barleycorns laid end to end. The original yard did not equal 3 times 36 barleycorns but was the measurement of the distance from the tip of King Henry I's nose to the tip of his outstretched arm extended sideways at the shoulder. The traditional mile was a carryover from the old Roman thousand paces[1] or the distance a marching legion could travel in 2,000 steps. Not even the Romans knew exactly how far this was. During the sixteenth century Queen Elizabeth I declared that the mile should be 5,280 feet, which would make the mile exactly 8 furlongs.

THE METER

The meter of the metric system evolved through the years, but much less dramatically than did the English linear units. Attempts at establishing forerunners of the meter that ended in failure included making the standard unit as the length of 1 minute of arc of the earth's meridian. Another proposal was to fix the standard unit as the length of a pendulum beating seconds. However it was in the midst of the French Revolution in 1790 that the National Assembly of France requested the French Academy of Sciences to prepare invariable standards for all measures and weights.[2] The meter was to be constructed so that it would equal one ten-millionth of the distance from the north pole to the equator

[1] A Roman pace, like the geometrical or great pace, is represented by the distance from the place where either foot is taken up, in walking, to that where the same foot is set down.
[2] Susan Fraker Holt: The United States and the metric system (Volume XI of Ninth District Exponent), Federal Reserve Bank of Minneapolis, 1973, Minneapolis, p. 3.

along the meridian of the earth running near Dunkirk in France and Barcelona in Spain. Today the length of the meter is defined as 1,650,763.73 wavelengths of the orange-red line of the spectrum of krypton 86 in a vacuum.[3] The International Bureau of Weights and Measures located at Sèvres, France, serves as the permanent coordinator to preserve and promote accurate measures.

Measuring length

In previous years, if a child were asked how tall he was in centimeters, it was most appropriate for him to multiply 2.54 times his height in inches. This obviously will give the correct answer but demands dependence on the English system. However, as the change to the metric system takes place, the goal becomes more than just getting the correct answer. Children must become familiar and at ease with the new system. For this to take place the idea of conversion must be discouraged and measuring devices marked in millimeters, centimeters, and meters must be provided. Even though the conversion approach is discouraged, no attempt should be made to prevent the child from discovering the relationship between the units of inches and centimeters. However, the emphasis of past years of first giving the student the conversion factor and a group of pencil paper problems should be avoided.

Teaching a child to become accustomed to and at ease with the linear measurements of the metric system involves active participation in measuring. One basic procedure is to provide the opportunity for each child to prepare his own measuring device. This gives him an additional opportunity to see and touch the various linear measurements. The local lumberyard can furnish white-pine lattice, which is inexpensive and can be made to approximate the width and thickness of a common ruler or yardstick. If the permanency of wood is not desired, heavy poster board cut in strips with the paper cutter provides an excellent stock for making rulers. For use as a standard in preparing their linear measuring devices, children should be provided commercially made rulers marked in millimeters or centimeters depending on the desired accuracy. The results of a class discussion may be used to determine the most appropriate length of the metric ruler. Those children who have had considerable exposure to the foot ruler will normally vote for a 30-centimeter ruler, because of its being very close to 12 inches in length. At an early age, children have stereotyped opinions as to the length a ruler should be from their past experiences.

After the measuring device has been prepared, the newcomer to the metric system is now ready to start measuring common objects in the surrounding environment. The teacher may wish to provide a list of items for pairs of students to measure, with one student being designated as "recorder" and one as "measurer." In affording each child the opportunity to measure, the teacher may give the assignment specifying rotation of duties at the midpoint of the activity. An example of such an assignment sheet is given in activity L-1 (p. 8).

In an attempt to learn how intermediate-grade children would attack a metric problem where the actual measurement would be difficult, an investigator posed the following problem to a sixth-grade class that had worked very little with the metric units of length:

---

[3] National Bureau of Standards: Brief history of measurement systems (Special Publication 304 A), Washington, D.C., 1972, Government Printing Office, pp. 2-3.

Your uncle who lives in France has come to your house for a visit. One night your uncle states that he would like to go to a basketball game. During the game he asks you if you know the distance from the floor to the basketball rim. Since he can only understand metric units, you will have to give him the answer in meters and centimeters. How are you going to find the distance using the metric system?

The answers varied considerably, with most students not being able to provide a definite method of finding the answer. Some students said, "Get a meterstick and a ladder and measure the distance"; others attempted to use various conversion approaches that were based on the fact that the rim was 10 feet above the floor. Here is one typical conversion answer:

Step 1. Take a dictionary
Step 2. Look on the table of measures and find the metric system and compare the two systems together.

An ingenious method was the following:

Since a meterstick is 3 inches longer than a yardstick, you would take 39 inches into 120 inches. This would be a little over 3 meters.

Problems such as these provide the opportunity for class discussion centered around alternative ways of solving linear problems. Even though children should be encouraged to find various ways of solving distance problems, the teacher should encourage actual measurement for those children who do not conceptualize the various linear units.

Children must become skillful in utilizing the prefixes of the metric system, which are used as multiples or submultiples of 10. One can best accomplish this by first providing the opportunity for children to work with the physical lengths of meter, centimeter, millimeter, etc. before launching a memorization activity to learn the table of prefixes. The teacher must determine when a child is ready to learn the prefix table. Most children can easily understand the significance of the metric prefixes during the second grade when they are able to skillfully count by tens.

Various games and activities that afford the learner opportunities to reinforce the initial concepts of the prefixes and how they act as multiplier and submultiplier of the basic unit will be found in this chapter. Added insight into the system can be presented by reminding the children that our number system like the metric system is based on powers of 10. The following comparisons may be helpful to the child:

Kilometer	Hectometer	Dekameter	Meter	Decimeter	Centimeter	Millimeter
km	hm	dam	m	dm	cm	mm
1,000	100	10	1	1/10	1/100	1/1,000
10^3	10^2	10^1	10^0	10^{-1}	10^{-2}	10^{-3}
Thousands	Hundreds	Tens	Ones	Tenths	Hundredths	Thousandths

Estimation

Even though developing the skill of estimation is a desirable goal, children have traditionally had little opportunity to acquire this ability. Exercises in estimation should receive a distinct emphasis, but only after considerable familiarization has been gained from actual measurement activities utilizing the units needed for the estimations, such as centimeters and meters. In the initial estimation activities, the children should concentrate on readily available small objects before attempting harder estimations. An activity that attempts to achieve this competency (L-2) is found later in this chapter.

If children are to realize the full advantages of the metric system, they must learn how to utilize decimals earlier than commonly taught. Even though there is some variation in sequencing by various textbooks series, decimals are normally introduced in the fifth grade. If a child is not able to take advantage of the greatest asset of the metric system, multiplying and dividing by powers of 10 by moving the decimal point, he is losing much of the value of the new system.

Diagnosis

Determination of whether a child should continue working with concrete measurement activities or whether he is ready to move ahead with semiconcrete or abstract work can be readily determined by informal procedures. Children may be asked to place their two index fingers on their desk so that they are 1 centimeter apart or draw a line 1 centimeter long. In either case it is possible for the teacher to survey the class to determine which children are able to conceptualize the centimeter length. This may be repeated for the decimeter. In order to demonstrate the length of a meter, the instructions may be given to hold out two fingers so that they are 1 meter apart. After children have conceptualized the length of the various metric units, they are ready to move to semiconcrete and abstract activities.

The following activities are designed to provide metric awareness of linear measurements by placing the emphasis on (1) actual measurement of objects and (2) pleasant game-like activities. These activities will provide a springboard to other teacher- and child-developed metric activities.

LENGTH ACTIVITIES

L-1. Metric measure

Purpose: To provide experiences measuring common objects

Materials: 3-decimeter rule

Number of participants: Any number of teams of 2

Directions: Choose a partner to assist in measuring the following items within the classroom. Fill in the names of the recorder and measurer. Change duties after making the first five measurements.

a. Recorder: ______________ Measurer: ______________

Length of math book ______________

Width of math book ______________

Height of recorder ______________

Length of teacher's desk ______________

Width of door ______________

b. Recorder: ____________________ Measurer: ____________________

Length of spelling book	________
Width of spelling book	________
Height of recorder	________
Width of teacher's desk	________
Width of window	________

L-2. Estimation

Purpose: To provide estimating experiences using mm and cm

Materials: Metric ruler

Number of participants: Any number of teams of 2

Directions: Divide into teams of two and estimate the size of each of the following objects without using your rule. After you have made your estimates, measure the objects and record the exact measurement beside your estimated value.

	Estimated	*Actual*
Length of a piece of new chalk	________ mm	________ mm
Longest distance across a penny	________ mm	________ mm
Width of your hand	________ mm	________ mm
Width of your desk top	________ cm	________ cm
Height of your teacher	________ cm	________ cm
Height of rim on basketball goal	________ cm	________ cm
Length of classroom	________ m	________ m
Distance from fourth- to fifth-grade door	________ m	________ m
Distance from floor to ceiling	________ m	________ m

Activities L-3, L-4, L-5, and L-6 are companion activities designed to utilize the materials described below. These materials may be cut out and assembled by the children and placed in individual envelopes marked METRIC KIT.

a. 10 1-centimeter lengths of drinking straw
b. 10 1-decimeter lengths of drinking straw
c. One 1-meter length of string
d. One card with "centimeter," one card with "decimeter," and one card with "meter"
e. One card with "cm," one card with "dm," and one card with "m."

L-3. Basic length activity

Purpose: To provide concrete experience with centimeter, decimeter, and meter measures

Materials: Metric kit

Number of participants: 1 to 30

Directions:

a. Ask the children to take from their envelope one of the shortest lengths of straw. Have them hold up this length of straw. While explaining to the group that this length is called a centimeter, record the new word on the board. Have the children reach into the envelope for the card that has "centimeter" printed on it. Discuss the fact that there is a short, easy way to abbreviate this long word. Have them find the cm card and place it beside the centimeter length of straw and centimeter card.

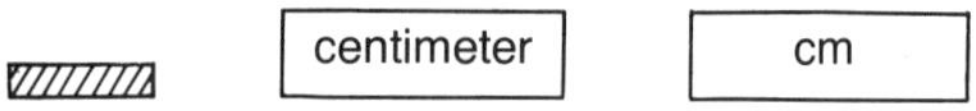

centimeter | cm

Ask them to place the materials on their desk as shown.

b. Repeat the above process by using one of the longest lengths of straw.
c. Ask the children to remove the long string from the envelope. Ask them to stand and let the string hang beside them. Explain that this length of string is a meter and may be abbreviated *m*.

L-4. Meter stretch 1

Purpose: To provide concrete experiences with metric length

Materials: Metric kits and die taped to cover spots, with "cm" written on three sides and "dm" written on three sides

Number of participants: 2 to 4

Directions: Empty the contents of several envelopes in a pile on the floor in easy reach of each player. Each player should tape on the floor a 1-meter length of string after it is fully stretched. Play moves in a clockwise direction with the player rolling the die to determine whether he gets to lay a cm length or a dm length of straw parallel to his meter length of string. Play continues until a winner has enough straw, end to end, to at least equal the meter length of string. (Cards may be substituted for the die by placing cm on 20 cards and dm on 20 cards. Shuffle the cards and place them face down. Each player takes a card in place of rolling the die.)

L-5. Meter stretch 2

Purpose: To provide concrete experiences with metric length

Materials: Same as L-4, except alter the die by placing "cm" on two sides of the die, "dm" on two sides, "no play" on one side, and "minus dm" on one side.

Number of participants: 2 to 4

Directions: Play continues the same as L-4 except the elements of "no gain" and "take away" have been added. Since going "in the hole" is not a possibility, the least a player can have is zero. (Ask children for more variations of the game.)

L-6. "Simon says"

Purpose: To provide concrete experiences with metric length

Materials: Metric kits

Number of participants: Any number

Directions: Using supplies from the kit, have the players show what Simon says. EXAMPLE: Simon says, "Place on your desk 2 centimeters of straw." Any player who responds to a direction not preceded by "Simon says" is caught. Last player caught wins.

L-7. Cuisenaire game

Purpose: To provide concrete experiences with centimeter lengths of 1 to 6

Materials: A regular cubical die, ample amount of w, r, g, p, y, and d Cuisenaire rods, and meterstick

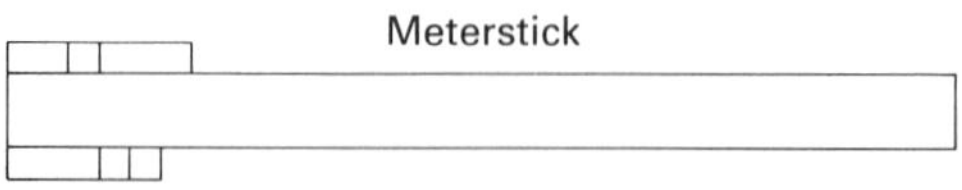

Meterstick

Number of participants: 2

Directions: Play moves in a clockwise direction with each player rolling the die to determine whether he gets to place a 1, 2, 3, 4, 5, or 6 centimeter length along his side of the meter-stick. Play continues until a winner has enough rods end to end to exactly equal 1 meter. (The game may be varied by preparing a direction die to accompany the previously described die. The direction die would have a "−" on one side, a "+" on four sides, and "lose a turn" on one side. If a "−" came up on one die and a 5 on another, the player would remove 5 cm of rod. Since going "in the hole" is not a possibility, the least that a player can have is zero.

L-8. 23 or block

Purpose: To provide concrete experiences with metric length

Materials: Cuisenaire rods—4 orange, 2 dark green, 2 yellow, 2 purple, 2 light green, 3 red, 3 white, 2 black

Number of participants: 2

Directions: Each player places end to end a train of rods consisting of 1 orange, 1 red, 1 orange, and 1 white.

orange	red	orange	white

Play begins by having each player in turn choose one rod from those remaining and placing it parallel to the train. A player receives points when he reaches the exact end of a Cuisenaire rod in the train. The points received are specified below.

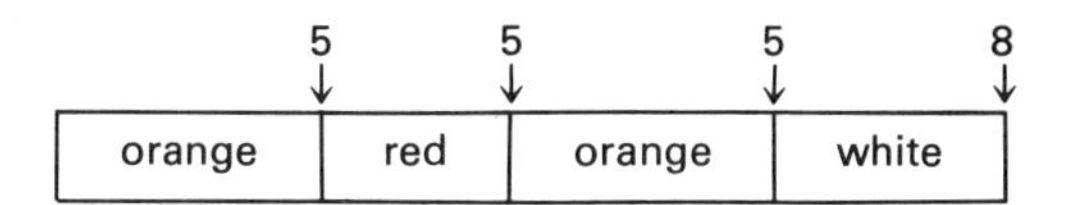

A player may choose the longest rods in an attempt to reach the end rod and score 8 points. Another player's strategy might be to score at each of the scoring points. The limited number of rods used in the game permits one player to block an opponent's scoring by selecting a needed rod. This blocking skill comes after playing several sets. Play continues until one player reaches the end of the white rod or both decide to stop. The player with the most points at the end of 4 sets wins the game.

L-9. Fill in the staircase

Purpose: To promote metric awareness of 1- to 10-centimeter lengths

Materials: A box of Cuisenaire rods and one die

Number of participants: 2

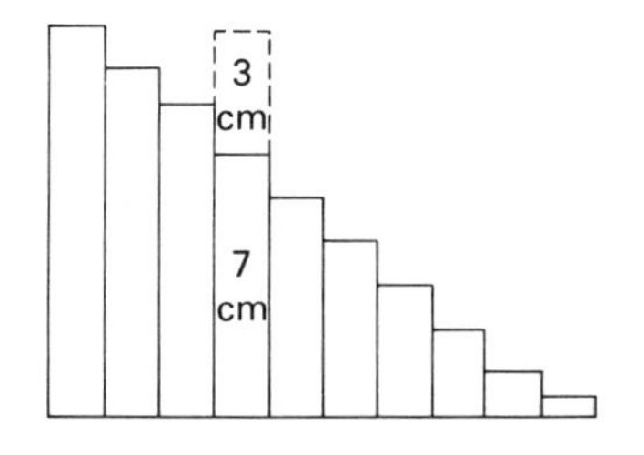

Directions: Each player arranges a staircase of 10 rods. Each then takes a turn rolling the die to determine what length from 1 to 6 centimeters he may use to help form a decimeter square of the staircase. If a player cannot use the designated length, he may not play during that turn. The first player to complete his square wins the game. The staircase and one play is shown in the diagram. Any combination of rods may be used to complete each decimeter length. Rods may only be played vertically.

L-10. 10 guesses

Purpose: To become familiar with metric linear terms

Materials: Objects in a room

Number of participants: 4 or 5

Directions: One child chooses something in the classroom that is in plain sight. If possible he may secretly measure the object. Then the players begin to ask questions to find out the dimensions of the object in terms of the metric system. Only 10 questions may be asked and all should involve metric length. The questions should be phrased so that they may be answered by a yes or no.

L-11. Meter master, may I?

Purpose: To become familiar with metric linear words

Materials: Squares of paper taped on the floor. On each square is written a different metric linear term, such as 100 cm, 2 dm, 34 m, 50 dam, etc.

Number of participants: 4

Directions: This game is a variation of the childhood game "Mother, may I?" One player is the "meter master." The other 3 players take turns being the "children." As indicated by the diagram, the meter master asks a child to move to a certain metric square on the first row, such as the 3 mm square. The child may move only if he says "Meter master, may I?" before he moves to the square. If the child does not say those words or does not move to the correct square designated by the meter master, he must go back to the starting line. The game proceeds one row at a move until a child reaches the last square to win the game and become the meter master. A diagram of the playing area is shown below.

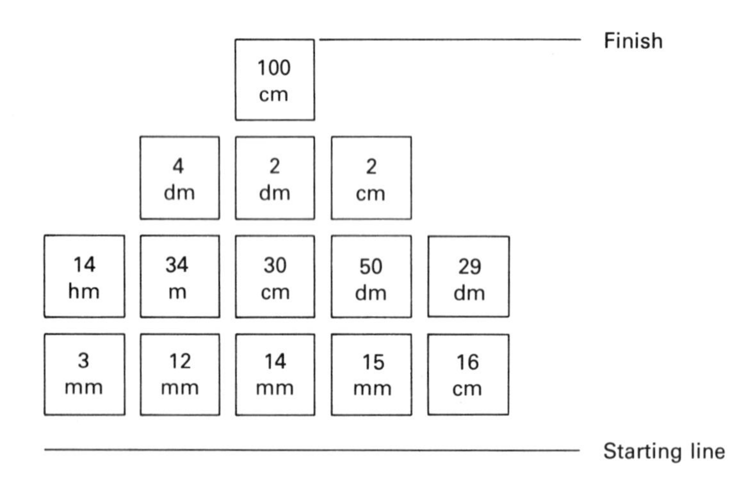

L-12. Chip trading

Purpose: To provide a game activity to acquaint children with conversion within the metric system

Materials: A chip till with the metric units taped on the till as shown in the diagram, one die, and an ample supply of chips

m	dm	cm	mm

Number of participants: Any number

Directions: Each player rolls the die to determine how many of the mm chips he may place on his till. When he has 10 mm chips, he may trade them for a cm chip. Play continues until a player wins by obtaining a m chip. To speed up the game, every third roll may be for cm chips.

L-13. Estimation contest

Purpose: To afford an opportunity for children to estimate lengths of concrete items after they have had considerable opportunity to become acquainted with metric units

Materials:

Lengths of string
- a. 1 meter
- b. 2 meters
- c. 5 meters
- d. 1 dekameter

Lengths of drinking straw
- a. 1 centimeter
- b. 2 centimeters
- c. 5 centimeters
- d. 1 decimeter
- e. 2 decimeters

Number of participants: Any number

Directions: Place these various lengths of straw and string at numbered stations around the classroom (the dekameter length of string may need to be placed in the hall). Children proceed individually from station to station, recording their estimate of the lengths. This activity may be set up by a pair of "judges." Added student interest may be obtained by providing badges for the judges and ribbons for the three best "estimators." One class may want to put on a contest for a neighboring class, whether younger, older, or the same age.

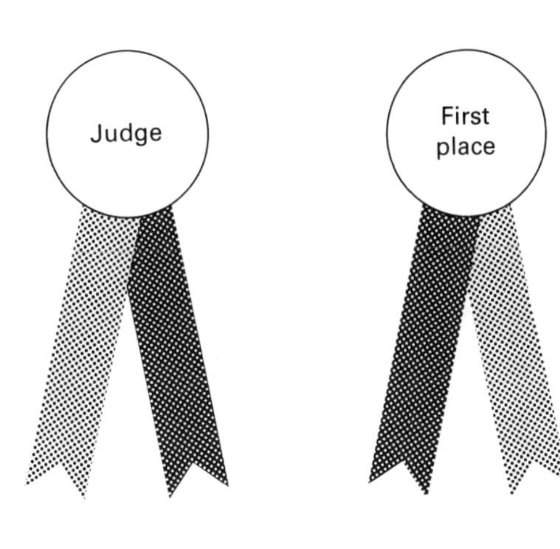

L-14. Looking for abbreviations

Purpose: To provide practice in recognition of linear abbreviations

Materials: Four abbreviation cards labeled mm, cm, dm, and m

Number of participants: 3

Directions: Give 2 cards to each of 2 children. Discuss with the children the abbreviation. Ask the children with cards to hold them up as the teacher says one of the linear terms. The child with no card tries to find the correct abbreviation for the pronounced word. If he finds the abbreviation, he reads it aloud. The child whose card has been read is now the "looker." Additional abbreviations may be added as needed.

L-15. Five-question secrets

Purpose: To provide an opportunity for children to verbally use metric linear terms

Materials: None

Number of participants: 5

Directions: One child chooses a linear word, such as centimeter, and whispers the name of it to another child. The remaining three children take turns asking questions and making guesses until one of them names the word or until 5 questions are asked. If the term is named before 5 questions are used, the 2 students who chose the term trade places with 2 other children in the class. If the term is not guessed, the 3 children trade places with other children in the class.

L-16. Which term is missing?

Purpose: To provide reinforcement for written metric words

Materials: Flannel board, metric linear word cards backed with flannel

Number of participants: Teams A and B with 5 on each team

Directions: Place 3 metric linear words on the flannel board. Have one of the children on team A close his eyes. Remove one of the words. Have the child open his eyes and tell which one is missing. Repeat the process with team B. Score one point for each correct answer. The number of words may be increased to make the game more difficult.

L-17. Linear scramble

Purpose: To provide a reinforcement activity with linear terms by individual students during free time

Materials: The following scrambled words

Number of participants: Any number

Directions: Unscramble the letters in each row to form a metric linear word. The letters in the △ (triangle) form a metric word.

imtmleelir _ _ △ _ _ _ _ _ _ _

ceemtired _ _ _ △ _ _ _ _ _

eniemcetrt _ _ △ _ _ _ _ _ _ _

mltekoire _ _ _ _ _ △ _ _ _

amkrdeeet _ _ _ △ _ _ _ _ _

temehoctre _ _ _ _ _ _ _ _ _ △

L-18. Pick a word

Purpose: To provide reinforcement for reading metric words

Materials: 20 cards with metric linear words on them

Number of participants: 3 or 4

Directions: Place the 20 cards face down in a pile. Have the children sit in a circle. Each child turns over a card when it is his turn. If the child can read the word, he keeps it. If he cannot, the next child may take that card or choose a new one. If he takes a new card, the unknown one is turned face down at the bottom of the pile. After all the cards have been used, the child with the most cards wins. Instead of using metric linear words you could substitute metric linear problems and follow the same directions. For example, you could have a problem such as, "How many meters in a hectometer," or "50 cm + 70 cm = __?__ meters." Ask the children to make up additional problems.

L-19. Fishing

Purpose: To provide linear problem-solving activities

Materials: The teacher or children may cut out fish from construction paper and put metric problems, metric linear abbreviations, or metric linear words on the fish.

Number of participants: 3

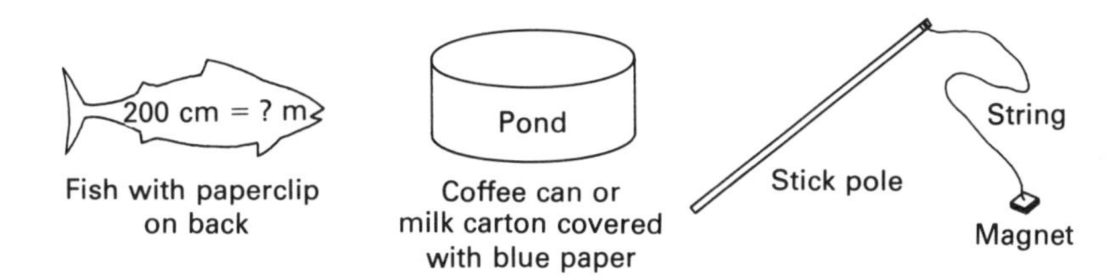

Directions: Put the fish in the pond and let the children take turns fishing. When a child catches a fish, let him solve the problem. If he solves the problem, he can keep the fish. The child with the most fish at the end of a given time wins the game.

L-20. Length concentration

Purpose: To provide for familiarization with the linear units and prefixes

Materials: 16 cards prepared so that each 2 cards have equal units except that they are written differently, such as 113 cm, 1.13 m; a separate set of small cards numbered 1 to 16

Number of participants: 2

Directions: Place the cards face down in 4 rows and 4 columns as shown in the example. During each player's turn, he is given the opportunity to call for turning over 2 cards. If these 2 cards are of equal value, he may hold them. If the 2 cards are not of equal value, they are returned face down. Play continues until all cards are held. The one who holds the most cards at the end is the winner. A new game may be started by reshuffling the cards and again placing them face down.

The example appears above.

1	2	3	4
5	6	7	8
9	10	11	12
13	14	15	16

L-21. Helping the coach

Before many official decisions that will change traditional measures such as the dimensions of the football field, height of the basketball goal, etc., are made, there will be excellent opportunities for student involvement in solving these problems.

Directions: Pose to a group of football-motivated students during the fall term the following problem:

You have been requested by the football coach to make recommendations to change the dimensions of the football field from English to metric units. What will be your recommendation?

Two solutions may be as follows:

a. Change the field to 90 meters, which is very close to the present 100 yards. Place the stripes at every 5 meters, which would permit a first down for each 10-meter gain. Of course, the midfield stripe would be on the 45-meter line.

b. The researchers may decide that their recommendation will be to lengthen the football field to 100 meters and then follow present rules.

L-22. Measure the teachers

Purpose: To provide experience in measuring and practice in graphing

Materials: Meterstick, graphing paper

Number of participants: 4 or 5

Directions: Estimate the height of the principal and other staff members. Ask politely if these persons would mind being measured the metric way. Divide this measuring activity among the members in your group. Record the information on a graph similar to the following:

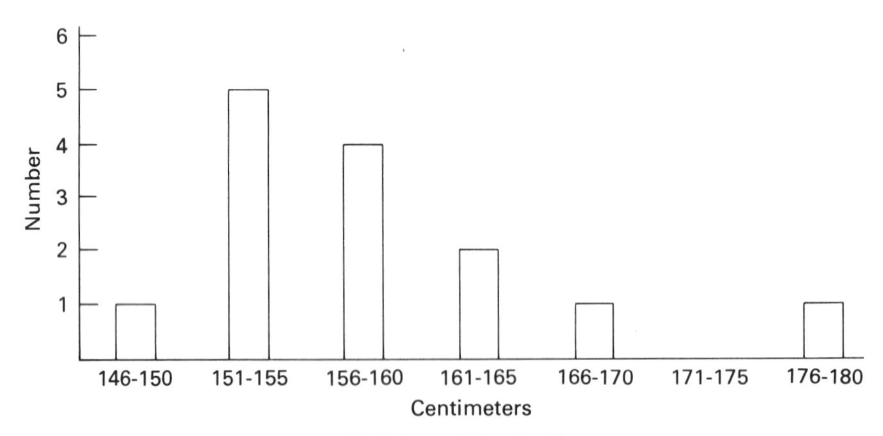

Heights of the staff at
Roosevelt Elementary School

L-23. Dimensions

Purpose: To familiarize children with measurements of common items in the classroom

Materials: Meterstick, paper, pencil, 3 × 5 index cards, and tape

Number of participants: Pairs of students

Directions: The teams of students are to measure the height and width of objects in the classroom. The teacher may make the assignments, or the children may volunteer to measure certain items. For example, 2 children may measure the chalkboard and put the correct height and length on a 3 × 5 index card. The students may use the cm, dm, and m units. Other teams could verify the data by performing the measurements themselves. If a mistake is found, they may make a challenge. The teacher will determine which team is correct. The cards should be left attached to the objects for a week or more to provide the children a chance to make the proper associations.

L-24. Trundle wheel push

Purpose: To learn to measure in meters using the trundle wheel

Materials: Trundle wheel

Number of participants: 3

Directions: Using the trundle wheel, have three children measure the following things:

a. Which is closer to the classroom, the boys' restroom or the girls' restroom?
b. How far is it to check out a good book?
c. How far is the distance a child must walk from his room to the place in the cafeteria where he picks up his tray?
d. Which is farther away, the gym or the principal's office?
e. Which play area on the playground is the biggest?

L-25. Metric cross numbers

Purpose: Provide abstract work with linear quantities

Materials: Work sheet

Number of participants: Any number

Directions: By following the arrows and operation, fill in the squares. For example, 50 mm + 50 mm is equal to 100 mm. When you are finished, the puzzle will check all directions.

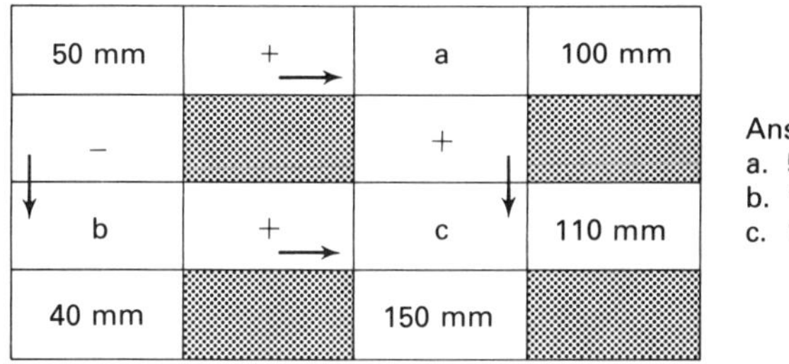

Answers:
a. 50 mm
b. 10 mm
c. 100 mm

The above activity may be made considerably more difficult by the following alteration:

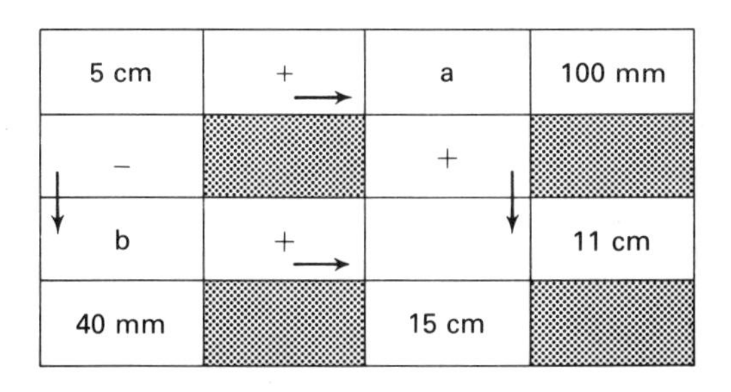

Answers:
a. 5 cm
b. 1 cm
c. 10 cm

L-26. Sum circle

Purpose: Provide practice with adding metric lengths (challenge activity)

Materials: Work sheet

Number of participants: Any number

Directions: Find the value of *a* and *b*. The sum of 2 numerals at the end of 2 diameters is equal to the sum of the 2 numerals on the opposite end of the same diameter. EXAMPLE: 40 mm + 20 mm = *a* + 1 mm. Therefore, *a* = 59 mm.

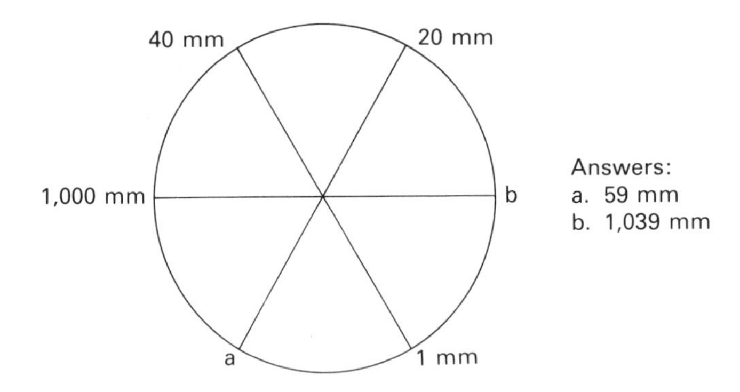

This problem can be made much more difficult by using the same values but different units as in the following:

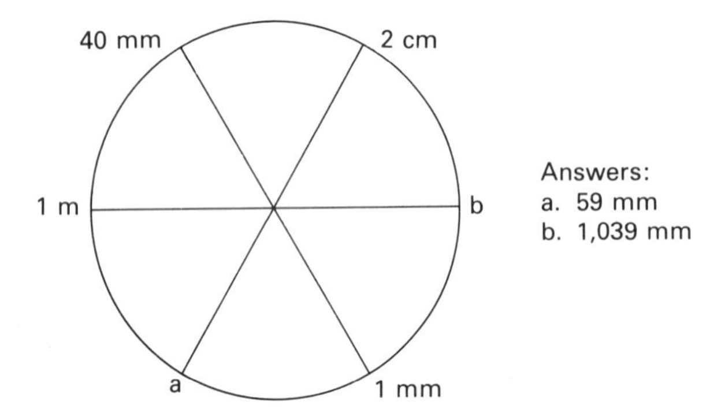

Ask the children to make sum circles for use in the class.

L-27. Equivalency jigsaw

Purpose: To practice the skills of renaming linear units

Materials: Puzzle constructed on poster board and cut on lines as shown in the picture; on each puzzle piece is a metric unit.

Number of participants: 1 or 2

Directions: Put the puzzle together by matching equivalencies.

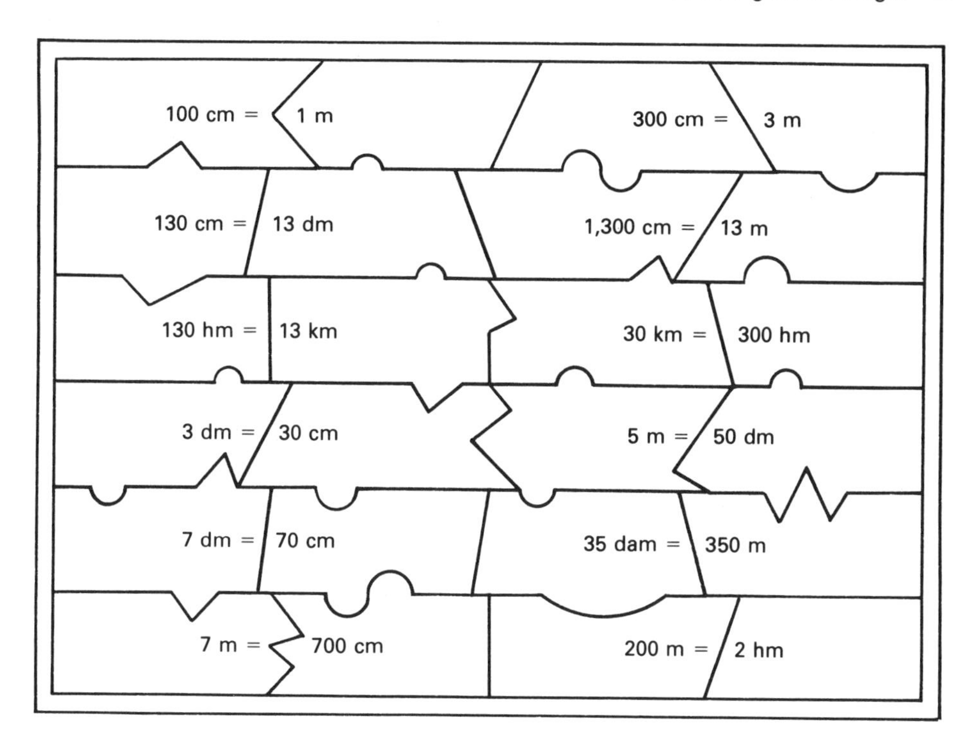

L-28. Pick-a-number quiz

Purpose: To provide re ... near units

Materials: Duplicated ... anying exercise

Number of participant

Directions: Underlin

How many

a. centimeters ... J 1000

b. meters in a ... 100 1000

c. meters in ... J 100 1000

d. hectomete ... : 10 100 1000

e. millimet... a meter? 10 100 1000

f. decimeters in a meter? 10 100 1000

g. millimeters in a decimeter? 10 100 1000

h. centimeters in a dekameter? 10 100 1000

i. dekameters in a kilometer? 10 100 1000

L-29. Supermetric puzzle

Purpose: To provide a challenge activity when the basic lesson is completed

Materials: A card containing the following problem

Number of participants: Any number

Directions: The teacher will help you check your answer when you have completed the problem.

A fish had a tail as long as its head plus 1/10 the length of its body. Its body is 9/10 of its total length. Its head was 10 cm long. What was the length of the fish? (This is difficult.)

Answer: Fish = 2,000 cm; head = 10 cm; body = 1,800 cm; tail = 190 cm

L-30. Loop a meter

Purpose: To provide reinforcement with an abstract activity

Materials: Accompanying chart

Number of participants: Any number

Directions: Draw a loop around each 3 boxes in a row (horizontally, vertically, or diagonally) that add to 1 meter.

40 cm	300 mm	3 dm	400 mm	2 dm	70 cm
3 dm	60 cm	500 mm	30 cm	2 dm	400 mm
500 mm	3 dm	40 cm	100 mm	8 dm	40 cm
10 cm	500 mm	4 dm	50 cm	300 mm	45 cm
1 dm	60 cm	200 mm	3 dm	150 mm	1 dm
300 mm	4 dm	20 cm	4 dm	20 cm	45 mm

L-31. Abstract conversions

Purpose: To provide experience converting within the metric system

Materials: Accompanying problems

Number of participants: Any number

Directions: Convert the following as indicated.

a. 1 m = ________ cm

b. 1 dm = ________ m

c. 1 mm = ________ cm

d. 1 km = ________ m

e. 1 hm	= ________	dam		k. 14 cm	= ________	dam
f. 1 dam	= ________	dm		l. 10 hm	= ________	km
g. 1 hm	= ________	cm		m. 1,000 mm	= ________	km
h. 1 km	= ________	dm		n. 100 cm	= ________	hm
i. 14 mm	= ________	cm		o. 3.4 mm	= ________	m
j. 25 cm	= ________	mm		p. 32.3 cm	= ________	dam

Answers:

a. 100	e. 10	i. 1.4	m. 0.001
b. 0.1	f. 100	j. 250	n. 0.01
c. 0.1	g. 10,000	k. 0.014	o. 0.0034
d. 1,000	h. 10,000	l. 1	p. 0.0323

CHAPTER 3

TEACHING METRIC WEIGHT

Weight and mass are terms often confused. This confusion is readily understood when one realizes that at the equator, at sea level, in a vacuum, weight and mass are equal and in other places on earth where they are not strictly equal they are still so close that it is possible to disregard the variation except in very precise scientific measurements. A brief discussion at the beginning of this chapter should clarify the proper use of each term.

Mass is a universal unit, which is the same everywhere, whereas weight varies according to its physical location. The mass of a specific person, such as 72 kilograms, would be the same as his weight on the equator at sea level. As this person moved from the equator, his weight would vary, whereas his mass would remain at 72 kilograms. The most commonly used example of this change is the comparison of the weight of an astronaut on the earth and on the moon. The 72-kilogram astronaut weighed at the earth's equator would weigh only about one sixth of this when weighed on the moon. When the space traveler lands on the moon, he finds that a jump will send him much farther than a jump of similar effort on earth. This change in weight is caused by the variation of gravitational pull of the moon as compared with the pull of the earth. During this variation in the astronaut's weight, his mass remains constant at 72 kilograms.

It is the purpose of these examples to exemplify that even though mass and weight for the same body are greatly different away from earth, on earth this difference is so small that for all except the most precise measurements, we can disregard it. Since this variation is so small, there seems to be no reason to unnecessarily complicate the topic for children by introducing the concept of mass. Obviously when the topic of space travel is discussed the preceding examples could be appropriately used. It appears evident that the term "weight" will continue to be used with the new metric units. An example of this is the present custom of using the term "weight" on food containers to indicate the measurement of the contents.

Having established that our units are applicable to weight or mass, let us discuss those units—the gram and the multiples and submultiples of it. Even though some are advocating teaching only milligram, gram, and kilogram, it is more consistent to teach children the 7 units, milligram, centigram, decigram, gram, dekagram, hectogram, and kilogram.

To teach only milligram, gram, and kilogram would be a variation in what already exists in place value, money, decimals, and previously taught metric units. To place the metric system on the same framework as already exists is to actually teach nothing new. It is true that in the practical world, the three most widely used units of weight will be the milligram, gram, and kilogram.

The decimal is like a period in a sentence. It ends the whole numbers, and the parts of the unit start at the right of the decimal point. The similarities of money, place value, and decimals are shown below:

Money	$1,000	$100	$10	$1	$0.10	$0.01	$0.001[1]
Metric	Kilogram	Hectogram	Dekagram	Gram	Decigram	Centigram	Milligram
Place value	Thousand	Hundred	Ten	One	Tenth	Hundredth	Thousandth
Decimal	1,000	100	10	1	0.1	0.01	0.001

With this approach it is much more consistent to treat the gram as the basic unit as was similarly done by using the meter as the basic unit of length. To treat the kilogram as the basic unit again diversifies the already existing sequence.[2] It is much easier to teach the pattern and then later diversify the already existing sequence. Each teacher of course may do what he or she believes best, but the paramount idea is that there should be consistency. Adopting the approach of teaching only the practical, commonly used units may bring on a confusion that could be easily prevented by teaching the pattern approach, which is common in the areas described. Even though, as previously stated, the most commonly used units are the milligram, gram, and kilogram, elementary school children will mainly use the gram and kilogram. The milligram, a small unit, is used extensively by pharmacists and medical workers and is commonly seen on vitamin bottles.

The largest metric unit that will commonly be used when metrication is fully realized is the metric ton. It will be used for weighing very heavy items such as cars, tractors, ships, etc. The metric ton consists of 1,000 kilograms. The metric ton is defined as the weight of water that is needed to fill a box 1 meter on each side.[3] The elementary child will have little contact with this unit, except in lessons where the exports of a country are given in metric tons.

Since a prime goal in teaching metric weight is to familiarize children with just how a gram or kilogram feels, they should have many opportunities to lift known weights. For this reason, each classroom must have marked weights that children may handle so as to gain an awareness of the metric weights. In addition to the various balance scales and weights available commercially, teachers may wish to prepare their own materials. An

[1] The mill is not a common unit of money but is used as a unit of taxation.

[2] Refer to Chapter 1 where the existing SI units are given and note that the unit for weight is the kilogram. However for simplification, uniformity, and understanding, I advocate in this book a patterned approach for initial instruction.

[3] The model of a cubic meter may be purchased commercially or it can be constructed from 12 metersticks. It is highly advantageous for children to actually see the size of the cubic meter.

economically constructed balance that weighs accurately is shown in the following drawing:

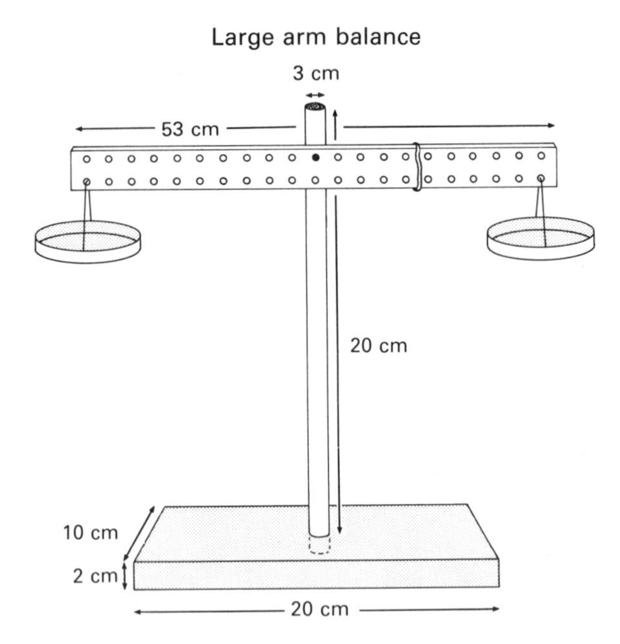

Materials: Balance arm is made from pegboard and pivots on a nail through the stand. Cups are of Styrofoam, approximately 2 cm deep. Stand has a large dowel rod approximately 3 cm in diameter. Base is any suitable piece of wood. Rubber band may be moved in or out to balance arm.

The basic equipment needed in each classroom for teaching metric weights of small magnitude consists of balance scales and a set of known weights. If one 1-gram, two 2-gram, and one 5-gram weights are available, objects from 1 to 10 grams may be weighed to the closest gram. Several activities that utilize prepared weights from either twist licorice or metal are presented later in this chapter. All activities presented are based on the concept that children must be given the opportunity to weigh objects if they are to become at ease with metric weights. Metric utilization will never be natural if a textbook, pencil, and paper approach is followed.

For children to have the experience of weighing themselves in kilograms, they must have appropriate scales for making such measurements. Appendix D lists companies where metric bathroom scales may be purchased. It is physically possible to replace the number scale on regular bathroom scales so they will measure in kilograms. However the availability of regular metric scales makes such an alteration unnecessary.

WEIGHT ACTIVITIES

W-1. Weighing licorice

Purpose: To prepare 1-, 2-, and 3-gram weights to be used in future activities

Materials: Balance scales, known weights, twist licorice, scissors

Number of participants: Any number of students working in pairs

Directions: After the teacher demonstrates how to use the balance scales, he or she will place a known 2-gram weight on one side of the scales and an equal weight of twist licorice on the other side. Licorice varies greatly, but this amount will be approximately 4 cm in length. Since the licorice must originally be heavier than 2 grams, the weigher may wish to eat the small amounts that are snipped from the licorice until it balances. Each student should prepare the three weights listed above for use in the following activities.

W-2. Graham-cracker snack

Purpose: To provide practice in weighing 6-gram amounts

Materials: Balance scales, licorice weights from W-1, and graham crackers

Number of participants: Any number of students working in pairs

Directions: Each student is to eat exactly 6 grams of graham cracker. This may be accomplished by snipping or breaking from one cracker until the amount remaining balances the combined 1-, 2-, and 3-gram pieces of licorice.

W-3. M & M weigh in

Purpose: To become familiar with 3-gram amounts

Materials: Balance scales, known licorice weights, and M & M Plain Chocolate Candies

Number of participants: Any number of students working in pairs

Directions: Each student may eat 3 grams of M & M candies after weighing them on the balance scales. How much does an M & M weigh?

W-4. Five grams of raisins

Purpose: To become familiar with 5-gram amounts

Materials: Balance scales, known licorice weights, and raisins

Number of participants: Any number of students working in pairs

Directions: Each student may eat 5 grams of raisins after weighing them on the balance scales. How much would 10 raisins weigh?

W-5. Preparation of permanent weights

Purpose: To provide those children who are in the fifth grade and older the opportunity to prepare permanent weights to be used during the school year

Materials: Balance scales, tin cans, tin snips or heavy scissors, accurate weights of 1, 2, and 5 grams, and number punches to mark weights

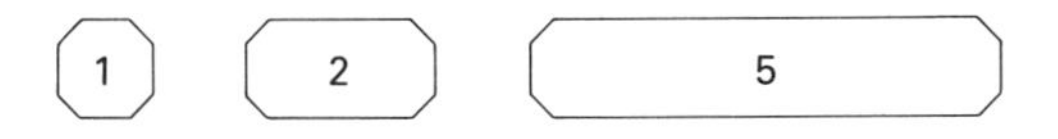

Number of participants: Any number of students working in pairs

Directions: Each team should cut from a tin can, an amount of metal a little heavier than the known weight. Small amounts of material should be snipped from the tin until the scales indicate that the amounts are equal. In order to weigh up to 10 grams to the nearest gram, the set of weights needed would be one of 1 gram, two of 2 grams, and one of 5 grams. The numerical amount may be scratched on the metal piece, or if number punches are available, the number of grams may be punched into the weight. During the process of snipping, the sharp corner of the weights should be trimmed. The weights are not considered dangerous if proper precautions are given as to the sharpness of the metal edges. Children enjoy the experience of making the permanent weights. Vinyl floor tile may be used with younger children if the danger of being cut on the metal is considered to be too great.

W-6. Who has the heaviest nickel?

Purpose: Gain experience in using balance scales

Materials: Balance scales, known weights and nickels

Number of participants: Any number

Directions: Announce to the class that tomorrow everyone should bring a nickel to be used in a metric activity. The activity is divided into two phases, with phase 1 being to determine if a heaviest nickel may be found and phase 2 involving the weighing of each nickel to the closest gram. Phase 1 may be carried out by placing one nickel opposite another on the balance scales.

W-7. What does a paper clip weigh?

Purpose: To acquaint children with metric weights

Materials: Balance scales, known weights, and paper clips

Number of participants: Any number of pairs of students

Directions: Ask each team to determine the different weights of 10, 6, and 2 paper clips. (Contrary to a common belief, many paper clips do not weigh 1 gram.)

W-8. Tang weigh in

Purpose: To provide experience in weighing gram amounts

Materials: Tang, glass, balance scales, known weights, water, and milliliter measurer

Number of participants: Any number of students working in pairs

Directions: Use the balance scales to determine the grams of powdered Tang that should be added to 125 ml of water to make a tasty drink. Start with 10 grams. (One-half glass of water may be used in place of the 125 ml of water.)

W-9. Chalk weight

Purpose: To gain experience in estimating gram amounts

Materials: Chalk, balance scales, and known weights

Number of participants: Any number

Directions: Weigh a new length of chalk before it is used. After the chalk is used for a day or two, estimate and record its remaining weight in grams. After the estimation is performed, weigh the chalk to determine its weight to the nearest gram. (A similar activity involves weighing a new pencil and then weighing it again after a week to determine its loss in weight to the nearest gram.)

W-10. Metric weight of students

Purpose: To establish each child's weight in kilograms

Materials: Metric bathroom scales

Number of participants: Any number of students working in pairs

Directions: After weighing, each student should record his weight for the following activity. (This activity may be inappropriate if a class member is sensitive to being excessively overweight.)

W-11. Graph of class weights

Purpose: To utilize data from W-10 to prepare a graph to the nearest 5 kg the weight of members of the class

Materials: Same as W-10

Number of participants: Total class

Directions: A committee should be selected to prepare a graph of the weights of the members of the class. An example of the graph of a class is as follows:

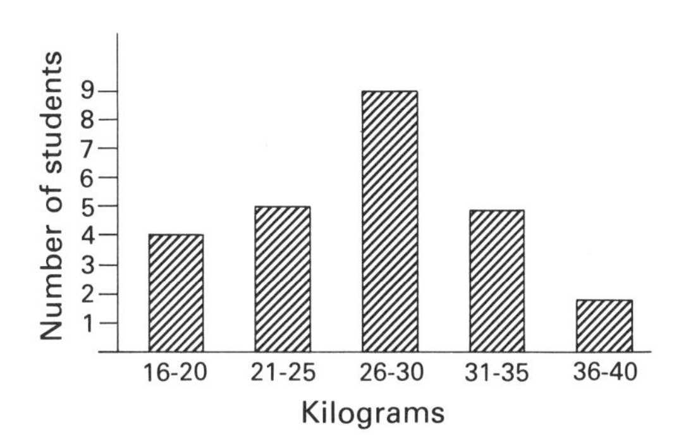

W-12. Verification of stated weight of food

Purpose: To acquaint children with the use of metric weights in measuring food

Materials: Balance scales, known weights, and some food that has the weight in grams recorded on the container, such as breakfast cereal and candy bars

Number of participants: Total class divided into teams of 4 or 5 students

Directions: To verify the new weight, one must weigh the food separately from the container. If the balance scales will not accommodate the bulky cereal in one weighing, it may need to be divided. Ask the children for alternative ways of determining the weight of food in a container without removing it from the container.

Answer: Obtain one empty bean can and an identical can that is full. Two weighings and a subtraction problem will divulge the new weight.

W-13. Heaviest and lightest person in school

Purpose: To provide children an opportunity to become aware of expressing the weights of persons in metric units; to provide total school awareness of metric measurements

Materials: Metric bathroom scales and material for badges

Number of participants: One committee of 4 to find the lightest person in school and one committee of 4 to find the heaviest person in school. A third committee of 4 may prepare the badges to be worn by the lightest and heaviest persons.

Directions: Each committee should ask permission to weigh a person they suspect to be the lightest or heaviest. Under no circumstances should a person be weighed unless he indicates an interest in the contest. Probably the heaviest person will be a male teacher and the lightest person will be a kindergarten child. The badges to be awarded may be similar to the following:

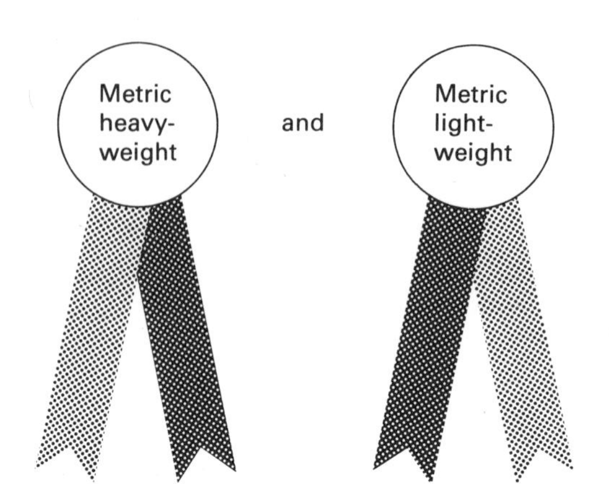

W-14. Metric food labels

Purpose: For children to discover the numerous foods that are being weighed in metric units

Materials: Food containers with metric weights listed

Number of participants: Total class divided into teams of 4 or 5 students

Directions: Have a contest to see which team can bring to school the most different food containers that have the weight recorded in metric units.

W-15. Make a kilogram

Purpose: To provide experience in making metric equivalencies within the metric system

Materials: 60 cards with varied amounts of grams less than 1,000 written on them, such as 300, 400, 325, 75, and 750 (Use only multiples of 25. Make several cards of 25, 50, and 100 g.)

Number of participants: 2 to 6

Directions: This game is played much like most common card games. Play begins with the dealer dealing seven cards to each player and then placing one card face up and the remaining cards in a stack face down. Each player takes his turn picking up a card and discarding another. He may pick up the top card from either the discard (face-up) pile or the face-down pile before discarding an unwanted card on the face-up pile. When a player holds sufficient cards that add to 1 kilogram, he may place them on the table close to his playing area. Play continues until all cards have been drawn or a player is able to lay down all his cards. The winner is the first person to reach 10 kg. This may take several hands of play. (The game may be made more challenging by subtracting what is held when play is stopped.)

W-16. Ten questions

Purpose: To provide weight familiarization in a game-like situation

Materials: 15 cards with whole units of metric weight written on them, such as 3 kg and 9 dag (No card can have more than 9 units.)

Number of participants: Total class

Directions: The teacher or a child draws a card but does not show it to the class. The remaining class members ask questions that may be answered by yes or no. For example, a good question might be, "Is it smaller than a kilogram?" The class wins if the answer is guessed before 10 questions are asked. (After playing as a class exercise as explained above, let the members play in pairs.)

W-17. Metric baseball

Purpose: To provide practice in stating equivalent terms

Materials: Cards on which problems for equivalent conversions are written. These conversions should vary from easy to difficult and will be marked single, double, triple, or home run. A single base hit would involve a question such as, "How many kilograms in 2,000 grams?" whereas a triple might be, "How many kilograms in 40,000 milligrams?"

Number of participants: Teams of 4 to 10

Directions: Place in the room the four bases of home, first, second, and third. The teacher or a designated umpire may oversee the game by determining if the answer is correct or if too much time has been taken. Each team gets 3 outs. Each base runner may advance as many bases as the batter advances on a hit. Each person gets one question with one answer while at bat. (A 30-second time limit is suggested for each question.)

W-18. Weight bingo

Purpose: To provide practice in recognizing metric abbreviations

Materials: Set of cards with abbreviations placed in a random manner so that each card is different; markers of some type such as corn or small pieces of paper

Number of participants: Total class

Directions: The teacher or a student may act as caller by calling out the metric term. There are various ways of winning, such as only a vertical line of three, only all four corners, and black out. This game can be made much more difficult by reading equivalent terms such as 100 g for a hg.

kg	mg	g
dag	Free	hg
cg	dg	30 mg

W-19. Weight circle scramble

Purpose: To provide a small group the opportunity to match equivalent terms

Materials: Cards with equivalent weights, such as 1 kg and 1,000 g, 1 dag and 10 g, 1,000 mg and 1 g, and 1 cg and 10 mg

Number of participants: 8

Directions: Players form a circle and hold up the weight cards. The teacher may say, "Scramble." This is the cue for the children to pair up in metric equivalents. The cards may be exchanged before another game is played.

W-20. Which expression is sick?

Purpose: To provide practice in expressing metric weights

Materials: Prepare 20 cards (some correct and some incorrect) that show stretched-out metric expressions, such as the following:

321 g = 3 hg + 2 dag + 1 g

4.25 g = 4 g + 2 dg + 5 cg
3.21 dag = 3 dag + 2 dg + 1 cg (incorrect)

Number of participants: 1

Directions: Sort the cards into a correct pile and an incorrect pile. (A stopwatch may be used if a contest situation is desired. The cards may be used in a relay game by having an equal number of players on each of two teams.)

W-21. Relay

Purpose: To provide practice in working with metric expressions

Materials: 20 cards each containing one weight expression larger than a gram, for example:

Number of participants: 2 teams of 5 to 10 each

Directions: The game is similar to other relay games. The teams line up, and the first person takes a card and goes to the board and records the amount in grams. If a player experiences difficulty, the teacher may provide hints. The first team to finish correctly is the winner.

W-22. Weight equivalences

Purpose: To provide a manipulative game activity to acquaint children with weight conversion within the metric system

Materials: An ample supply of small squares of different colored paper with only one abbreviation dag, g, dg, cg, or mg written on each; a sheet with columns marked as in the accompanying diagram and 2 dice. One die is taped with mg, cg, and dg, each written on two sides.

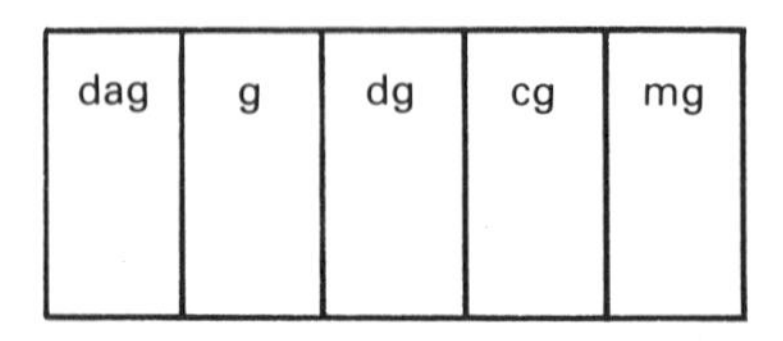

Number of participants: Any number

Directions: Play begins by a player rolling the 2 dice to determine how many markers he gets to place on the chart. If a 5 and a cg should come up on the dice, the player may put 5 cg markers on his cg column. Whenever a column has 10 markers in it, a trade should be made for a marker in the column to the left. The first person to gain a dag marker is the winner. It works well to play with three teams of 2 members each, with one member rolling the dice and one member recording.

W-23. Equivalency wheel

Purpose: To provide experience converting within the metric system

Materials: Playing board with spinner as shown in the example

Number of participants: Teams of students (may be played with 2 players)

Directions: Each team sends a member to the wheel to spin a metric weight. To obtain 3 points, the student must correctly verbalize the amount where the arrow stops and the amount stated in units above and below; for example, if the arrow stopped on 80 hg, the player must give 8 kg and 800 dag. He would receive only 1 point if he could only state the 80 hectograms.

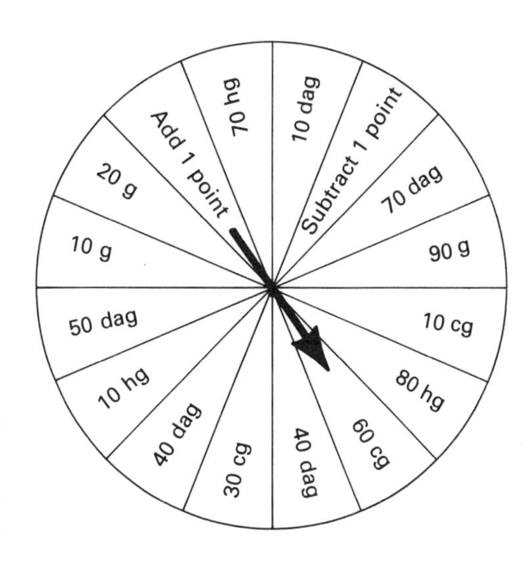

W-24. To tell the truth in metric weight

Purpose: To provide a class activity that provides metric-weight awareness

Materials: Three large cards on which are recorded, "I am a kilogram." The back of each card contains one of the following: "I weigh 10 dekagrams," and "I weigh 100 dekagrams," and "I weigh 500 grams."

Number of participants: 3 contestants, a panel of 4, and the remaining members of the class acting as the audience. The teacher is the moderator during the first few contests.

Directions: The 3 contestants sit in front of the panel with each saying, "I am a kilogram." Each contestant then gives one more clue such as, "I weigh 100 dekagrams," "I weigh 500 grams," and "I weigh 10 dekagrams." The panel then records the number of the person they think is the real kilogram and holds it up for the audience to see. The moderator then says, "Will the real kilogram please stand." The game may continue using different cards and exchanging the members of the panel and the contestants with members of the audience. (This is a favorite activity.)

W-25. Deal-a-gram (enrichment level)

Purpose: To provide experience in making metric equivalencies within the metric system

Materials: 54 cards—2 each of 1 to 9 dg, 2 each of 1 to 9 cg, and 2 each of 1 to 9 mg

Number of participants: 2 to 4

Directions: Deal 7 cards to each player. Place the remaining cards face down in a stack and turn the top card over. This becomes the discard stack. The player to the left of the dealer begins by either taking the top card in the discard stack or drawing a card from the reserve stack. He must return a card to the discard stack unless he can form books with all of his cards. The object of the game is to get as many books (2 cards or more) as possible that total to 1 gram, such as 9 dg, 9 cg, and 1 cg. Play stops when a player uses all his cards or all the cards have been taken from the reserve stack. The player with the most books wins the round. At the end of 4 rounds, the player with the most books wins the game.

W-26. Weight match

Purpose: To provide reinforcement of weight terms with a puzzle-like activity

Materials: Cardboard pizza disks, plastic clothespins, and felt-tip pen

Number of participants: Any number of students working individually or in pairs

Directions: Prepare the pizza disk as represented in the diagram. Place the 6 weight abbreviations of centigram, decigram, gram, dekagram, hectogram, and kilogram on plastic clothespins. Individual class members or teams may then place the clothespins in the correct spaces on the disk.

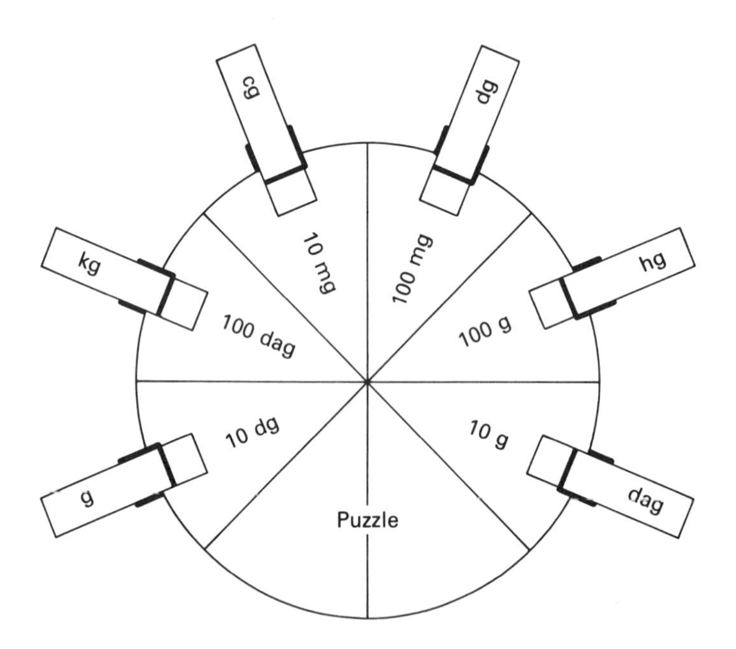

W-27. Gram-o-rummy

Purpose: To provide experience in learning the units of measure for the metric system from largest to smallest

Materials: 28 cards prepared with the following designations: 4 with mg, 4 with cg, 4 with dg, 4 with g, 4 with dag, 4 with hg, 4 with kg

Number of participants: 2 or 3

Directions: Deal 5 cards to each player. Place the remaining cards face down in a reserve stack and turn the top card over. This becomes the discard stack. The player to the left of the dealer begins by either taking the top card in the discard stack or drawing a card from the reserve stack. The object of the game is to make a run of 5 cards in consecutive order beginning at any point. Each player in turn, moving counterclockwise, either draws a card from the reserve stack or takes the top card from the discard stack. He discards a card from his hand each time he plays. The first player to get a run, such as kg, hg, dag, g, dg; or dag, g, dg, cg, mg, lays his cards down and he is declared the winner. (Children won't stop playing this game.)

W-28. Spin a kilogram

Purpose: Provides advanced activities in adding units of metric weight

Materials: Spinner made from a pizza board or the top of a cottage cheese container as follows:

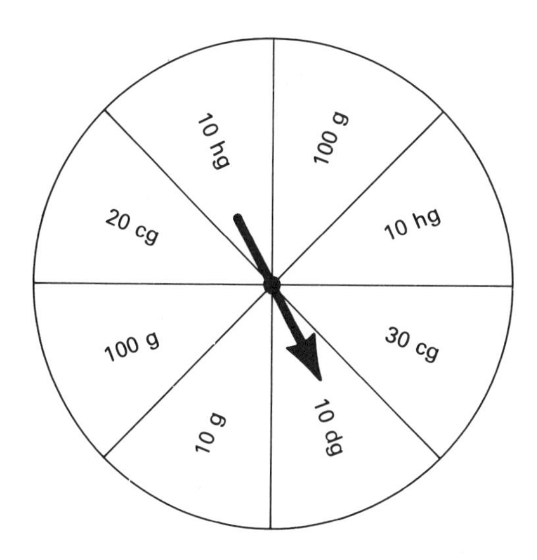

Number of participants: 2 to 4

Directions: Each player spins 2 times and adds the 2 weights. Play continues until someone reaches a total score of 1 kilogram.

W-29. Abstract conversions

Purpose: To provide experience converting within the metric system

Materials: Accompanying problems

Number of participants: Any number

Directions: Convert the following as indicated.

a. 1 g =	________ mg	i. 20.3 mg =	________ g
b. 1 cg =	________ mg	j. 21 dg =	________ cg
c. 1 kg =	________ g	k. 24 g =	________ mg
d. 100 g =	________ dag	l. 16 hg =	________ dag
e. 1,000 g =	________ hg	m. 0.01 kg =	________ g
f. 100 mg =	________ dg	n. 0.03 hg =	________ cg
g. 300 mg =	________ dag	o. 0.003 kg=	________ g
h. 3.2 cg =	________ g	p. 3.045 kg=	________ g

Answers:

a. 1,000	d. 10	g. 0.03	j. 210	m. 10	p. 3,045
b. 10	e. 10	h. 0.032	k. 24,000	n. 300	
c. 1,000	f. 1	i. 0.0203	l. 160	o. 3	

CHAPTER 4

TEACHING METRIC AREA AND VOLUME

When discussing the implementation of the metric system, it is common to hear teachers comment that perhaps measurement will now be better taught than it has been in the past. This optimism is attributable to (1) added emphasis in measurement because of the newness of the metric system and (2) the increased use of concrete activities to teach understanding of the new units. Many teachers have not been pleased with the measurement competencies that their children have possessed at the end of the school year. It is hoped that the optimism for better teaching of measurement concepts will become an actuality, in that children will have a better understanding of liters and square meters than they presently have of gallons and square yards.

AREA

Traditionally the concept of area has been introduced in the third or fourth grade with the measurement of area beginning in the fifth grade. There seems to be no reason to think that the change to the metric system will alter this sequence. If the findings of Piaget[1] are followed, area formulas should be delayed until fifth or sixth grade. Children cannot fully understand calculation of area until they achieve formal operational thought, and this does not occur for most children until they reach the age of 11 or 12. This means that readiness work for area should start in the third and fourth grade. The most appropriate readiness activities would be concrete ones that utilize graph paper, geoboards, and measuring devices such as the meterstick and trundle wheel.

The metric measures of area that will be most commonly used are given in the following manner:

Unit	*Symbol*	*Relationship*
Square centimeter	cm^2	$1\ cm^2 = 100\ mm^2$
Square decimeter	dm^2	$1\ dm^2 = 100\ cm^2$
Square meter	m^2	$1\ m^2 = 100\ dm^2$
Are	a	$1\ a = 1\ dam^2$
Hectare	ha	$1\ ha = 100\ a$
Square kilometer	km^2	$1\ km^2 = 100\ ha$

[1] Jean Piaget, Bärbel Inhelder, and Alina Szeminska: The child's conception of geometry, New York, 1960, Basic Books, Inc., p. 352.

Many interesting discussions are presently taking place, concerning just what metric units of area will be used to measure specific items. It seems quite logical that those objects that are measured by the square yard will be measured by the square meter, such as carpeting; those items that are measured by the square foot such as floor space will be measured by the square decimeter; and smaller items will be measured in square centimeters rather than square inches.

Even though it stretches the mind to think of the farmers and ranchers measuring their spreads in hectares, it appears that this method of measurement may not be too far into the future. However with the United States being surveyed by square miles, each of which consists of 640 acres, it is obvious that there are problems in the future that will not easily be resolved. One easy way out of this dilemma is to say that such a problem is beyond the scope of the elementary school student.

The activities that follow are designed not only to acquaint children with metric measures of area, but also to provide an understanding of the concept of area.

AREA ACTIVITIES

A-1. Form a square decimeter

Purpose: To acquaint children with square centimeters and square decimeters

Materials: Playing board 10 centimeters on a side, a die, and numerous pieces of paper in 6 sizes: 1 square centimeter to 6 square centimeters. (Playing board and area pieces may be made from tag board, as shown below.)

Number of participants: 2 to 4

Directions: Play begins by rolling the die to determine what metric area, such as 1 to 6, may be played. The winner is the first person to exactly fill the square decimeter. No piece may be moved after it is played, and a player must take the exact amount rolled each time in one piece.

A-2. What is the area?

Purpose: To provide an opportunity to find the area of items in the classroom

Materials: The 1-square-decimeter playing board used in A-1

Number of participants: Any number

Directions: Instruct the children to find the area of the top of their desk, the teacher's desk, etc. by using the square decimeter board. Answers may be given to the nearest whole square decimeter or in parts of a square decimeter.

A-3. Constructing a checkerboard

Purpose: To provide awareness of metric area

Materials: Tagboard, centimeter ruler, black and red liquid crayons, and a checkerboard to use as a model

Number of participants: Any number

Directions: Instruct the children to prepare a checkerboard that is measured in square centimeters. Since a square centimeter would be too small for checker placement, ask the children to determine what size they should use.

A-4. Foot area

Purpose: Provide concrete experience in working with metric areas

Materials: Graph paper marked in centimeters

Number of participants: Any number

Directions: Ask the question, "What is the area of your right foot in square centimeters?" After providing a period of time for discussion concerning how this measurement could be made, ask each student to draw around his foot on the graph paper. After this is done, count those complete square centimeters within the drawing and then estimate the partial squares. (This activity may be followed by determining how much pressure is put on each square centimeter when a child stands on one foot. This would be done by dividing the number of square centimeters in the foot into the child's weight in grams.)

A-5. Find the area

Purpose: Provide concrete experiences in finding areas

Materials: Cuisenaire rods, other blocks that have metric dimensions, or the playing pieces used in A-1 and the accompanying figure

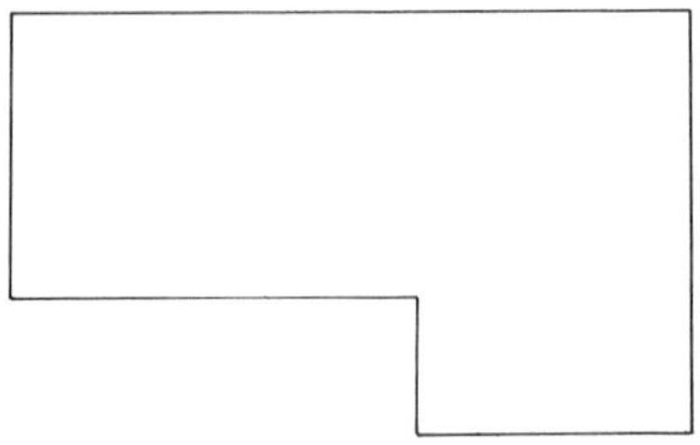

Number of participants: Any number

Directions: Determine the area of the following figure by filling the area with rods, blocks, or playing pieces. (For additional problems, ask the children to make different figures on index cards with the answers recorded on the back.)

A-6. Decimeter geoboard

Purpose: To provide concrete experiences with metric area

Materials: Geoboards marked off in centimeters with a nail at 2-centimeter intervals as illustrated in the diagram. The lines must be drawn on the board to show the square centimeter units. CONSTRUCTION HINTS: Children can construct these boards with guidance. The wooden base should be 12 cm × 12 cm. Blocks of this size are considered scrap at a lumber yard. The nails are small finish nails approximately 3.5 cm long. The lines should be made with a ball-point pen. Invite a couple of fathers to come and assist with this project. If this is impossible, ask some upper-grade boys to assist.

Number of participants: Any number

Directions: After demonstrating to the children that each square is equal to 1 square centimeter, ask the children to place a rubber band around 4 square centimeters and then around 8 square centimeters. Ask if any one can figure out different shapes that would include 4 square centimeters and 8 square centimeters. (Triangle shapes could be used.)

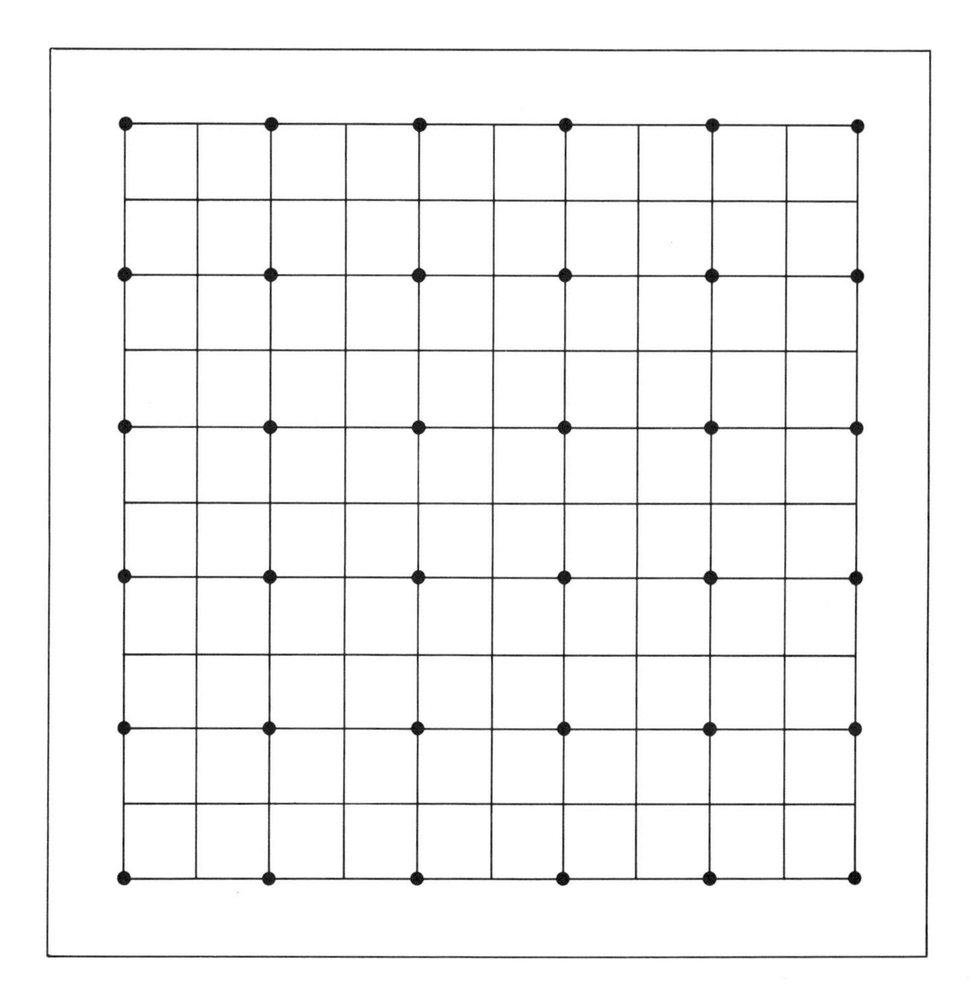

A-7. Geoboard activities

Purpose: To provide concrete experiences with metric area

Materials: Geoboards as used in A-6

Number of participants: Any number

Directions: Following are some things to do with a metric geoboard.

a. Construct the smallest square on the board.

b. Find the second largest square. What is its perimeter? What is its area? (It is not 16 square centimeters.)

c. Find the third largest square.

d. Construct a triangle of 2 square centimeters, of 4 square centimeters, of 6 square centimeters, and of 8 square centimeters. Did everyone use exactly the same shape? (Remember as long as the base and altitude remain the same, the area remains constant.)

e. Construct the floor plan of a house with the living room 8 square centimeters, kitchen 4 square centimeters, and two bedrooms 4 square centimeters each.

f. Construct a rectangle with area of 20 square centimeters, of 24 square centimeters, of 32 square centimeters, and of 28 square centimeters. (The last one cannot be constructed on this geoboard because of the spacings of the nails.)

g. Construct parallelograms having areas of 8 square centimeters and 24 square centimeters.

h. Construct trapezoids having areas of 8 square centimeters and 16 square centimeters.

i. Construct right triangles with areas of 2 square centimeters and 8 square centimeters. (Can others be constructed?)

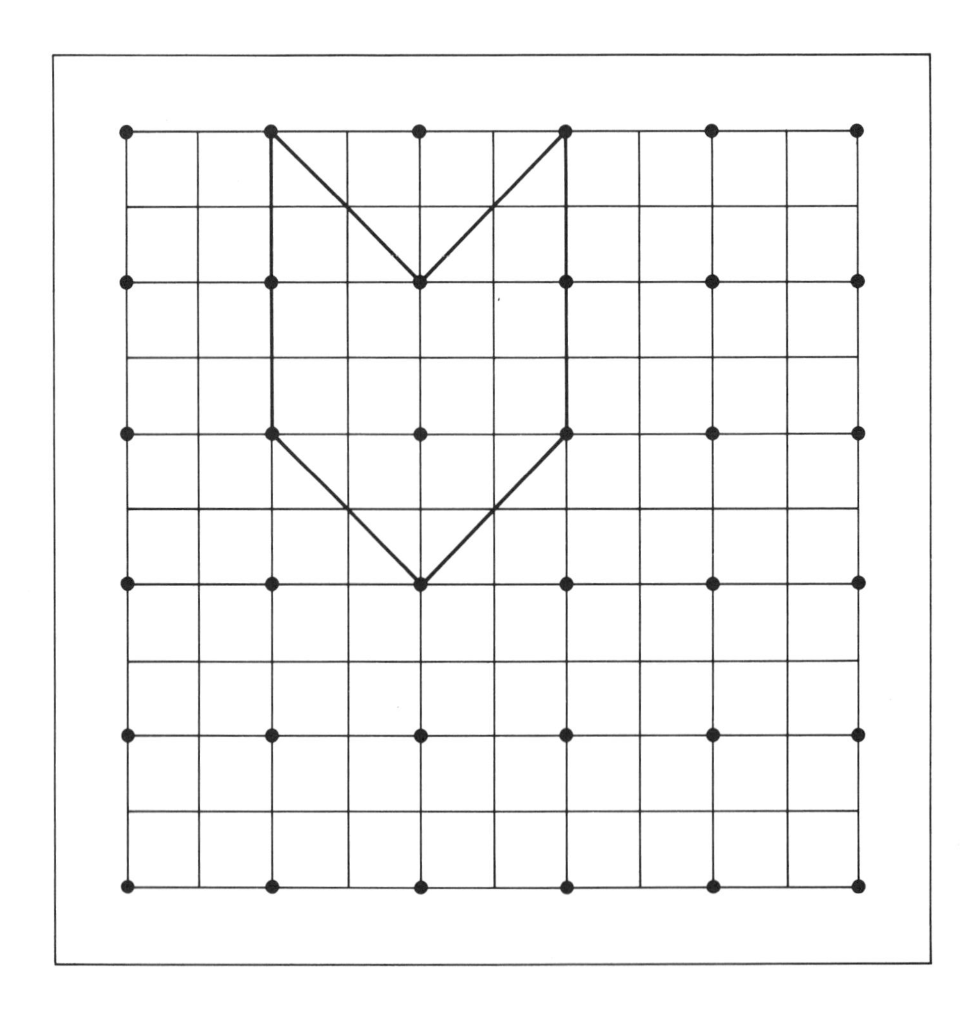

A-8. Geoboard activities

Purpose: To find areas without using a formula-solution process

Materials: Geoboards as used in A-6

Number of participants: Any number

Directions: Construct the previous figure and find its area. Make another figure and see if your neighbor can find its area.

A-9. Are the areas equal?

Purpose: To provide concrete experience with metric area

Materials: Geoboards as used in A-6

Number of participants: Any number

Directions: Construct the following 3 figures on your geoboard. Which of the figures have equal areas? Could you make another figure with equal area?

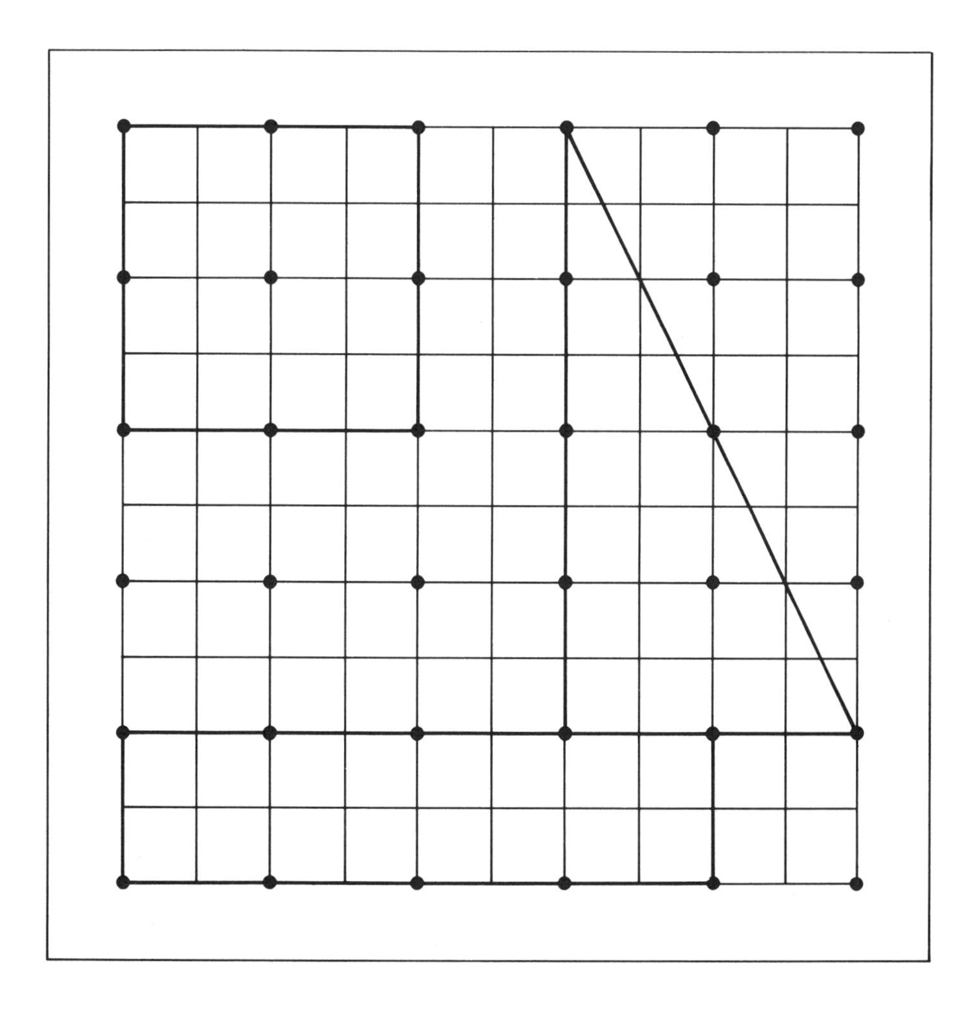

A-10. Metric farm

Purpose: To provide experience in working with ares and hectares

Materials: Pencil and paper

Number of participants: Total class

Directions: After explaining to the class that an are is the area of 1 square dekameter and that 100 ares make one hectare, show that a rectangular farm 1 kilometer wide and 2 kilometers long consists of 20,000 ares or 200 ha.

2 km

1 km

200 ha

Solution: First change both dimensions to meters and then multiply to find area in square meters. To change to ares, divide by 100 square meters, which is the number of square meters in an are. To change from square meters to hectares, divide by 10,000 square meters, which is the number of square meters in a hectare.

$$\text{Area} = 1{,}000 \text{ m} \times 2{,}000 \text{ m} = 2{,}000{,}000 \text{ m}^2$$

$$\text{Ares} = \frac{2{,}000{,}000 \text{ m}^2}{100 \text{ m}^2} = 20{,}000 \text{ ares}$$

$$\text{Hectares} = \frac{2{,}000{,}000 \text{ m}^2}{10{,}000 \text{ m}^2} = 200 \text{ ha}$$

Additional problems:

a. Find the number of hectares in a small farm 500 m by 1 km.

b. Find the number of hectares in a ranch 40 hm by 50 km.

Answers:

a. 50 ha

b. 20,000 ha

A-11. Area of school ground

Purpose: To provide an awareness of the are unit

Materials: A device for measuring long distances such as a trundle wheel, a surveyor's chain, or a cord that is premeasured to 10 meters

Number of participants: Teams of 4

Directions: Select an area that can easily be measured such as a portion of the school ground, a tennis court, the football field, or the gymnasium. After determining the dimensions, calculate the area in square meters. This figure should be divided by 100 square meters to determine the ares. (After the number of ares are determined, publish this by placing it on the school bulletin board, or in a metric newspaper that the class might publish for a few issues.)

A-12. Scaled area

Purpose: To provide experience in working with scaled drawings

Materials: Pencil and metric ruler

Number of participants: Any number

Directions: If an apple orchard is drawn to a scale such that 1 centimeter represents 1 dekameter, what is the area in ares of a rectangular orchard whose drawing is 5 cm by 3 cm? (Ask selected children to prepare additional problems for class use.)

A-13. Compute the area

Purpose: To provide experience in computing area

Materials: Accompanying figure, pencil, and metric ruler

Number of participants: Any number

Directions: Compute the area of the following figure. Use only a metric ruler, pencil, and paper. (For additional problems, ask the children to make different figures on index cards with the answers recorded on the back.)

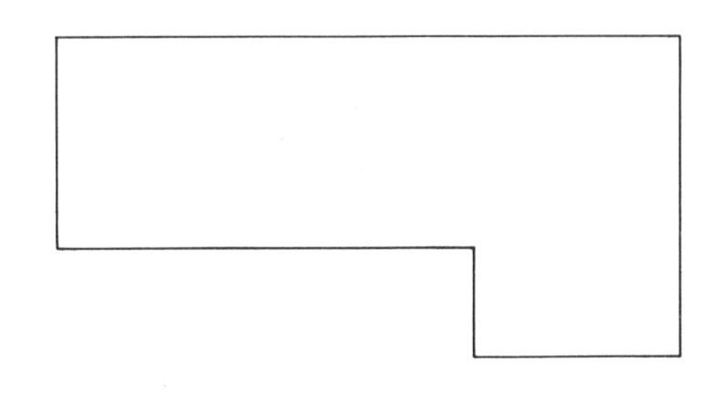

A-14. Gymnasium-scaled drawing

Purpose: To provide computational type of problems

Materials: Metric ruler, meterstick, pencil, and paper

Number of participants: Teams of 4

Directions: Using a scale such that 1 centimeter represents 5 meters, have children make a diagram of the gymnasium. Have several teams perform this activity before comparing resulting figures. Let children continue this activity by making other scaled drawings of the classroom, hallway, etc. Possibly a team would like to make a scaled drawing of the school building. Such activities should be encouraged.

VOLUME

The study of solid figures creates considerable difficulty for children. This is largely attributable to the attempt to exemplify 3-dimensional space figures on a flat surface such as a textbook page. It has been the design of too many elementary programs to overcome confusion by relying on frequent exposure to such drawings supplemented with opportunities for children to copy 3-dimensional figures from the text. That this approach has failed in far too many instances is evidenced by the memorized approach to solving volume problems, which leads to a lack of understanding.

Volume should be studied only after children have gained an understanding of perimeter and area of flat surfaces. Normally, readiness work for volume begins in the fourth and fifth grades, with formal work with formulas beginning in the sixth grade. The readiness activities must start with an intuitive awareness of a space region. This is accomplished by letting children handle shoe boxes, blocks, cans, and other common solid shapes. Children should be given an opportunity to stack 1-centimeter cubes in order to determine the volume of a small cube or rectangular solid. Blocks that are constructed with metric measurements are readily available at a reasonable cost. Every classroom should have these teaching aids available.

The relation of metric units of volume to each other provides a great advantage over the English system. This relationship is shown by the following:

1 cubic centimeter (cm^3) = 1 milliliter
1 cubic decimeter (dm^3) = 1 liter
1 cubic meter (m^3) = 1 kiloliter

One can see that metric volume consists essentially of the cubic centimeter, cubic decimeter, and cubic meter, as well as the liter and its multiples and fractions. In order to gain an understanding of these units, children need beginning activities that rely heavily on concrete measurements. Such activities permit the children to see and feel the volume rather than demanding that an abstract mental picture be formulated.

Those who work with older students and adults are aware that the present method of teaching cubic measurement has not been too effective. This is evidenced when students give the area of a floor in cubic units or the volume of a solid in square units. The activities of this chapter are designed to help alleviate this problem.

VOLUME ACTIVITIES

V-1. Preparation of a graduated cylinder

Purpose: To provide actual experiences with 50, 100, and 150 ml amounts of water

Materials: Graduated beaker or cylinder that will measure 50 ml amounts, small jar such as an olive jar, masking tape, permanent felt-tip marker (rubber bands may be used rather than the marker and masking tape)

Number of participants: Any number of teams of 2 children

Directions: Ask each team to prepare a graduated cylinder marked 50 ml, 100 ml, and 150 ml by using the vessel of known amounts and the small jar. The graduations for the above amounts may be marked on the masking tape with the felt-tip marker.

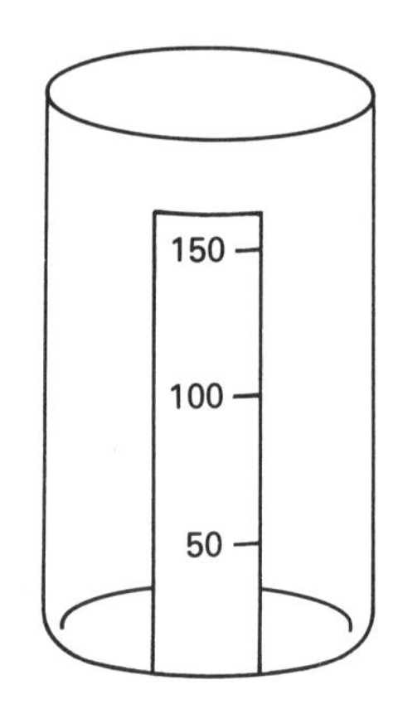

V-2. 100-milliliter swallow

Purpose: To provide actual experience with milliliter amounts

Materials: Graduated cylinder prepared in V-1

Number of participants: Any number

Directions: Measure 100 ml of water into the graduated cylinder. How many normal swallows does it take to drink the 100 ml of water?

V-3. Kool-Aid drink

Purpose: To provide actual experience with milliliter amounts

Materials: Kool-Aid as prepared in O-2 and the graduated cylinder prepared in V-1. Ask children to bring ingredients.

Number of participants: Any number of teams of 2 children

Directions: Sample the Kool-Aid to determine how many milliliters it takes to satisfy your thirst.

V-4. Volume of a baby food jar

Purpose: To provide metric awareness of milliliter amounts

Materials: Graduated cylinder or beaker marked in milliliters, baby food jar, and masking tape

Number of participants: Any number

Directions: Determine how many milliliters of water the baby food jar will hold if filled to the neck. Record this amount on masking tape placed on the jar.

V-5. Food container verification

Purpose: To provide metric awareness of milliliter amounts

Materials: Graduated cylinder or beaker marked in milliliters and a food container such as a fruit can with contents marked in milliliters

Number of participants: Any number of teams of 2 children

Directions: Using the graduated cylinder or beaker, verify that the marking on the food container is correct.

V-6. Milk carton measure

Purpose: To provide experience with milliliter measures

Materials: Small milk carton and beaker or cylinder graduated in milliliters

Number of participants: Any number

Directions: Fill an empty milk container with water and pour it into a graduated container to determine how many milliliters of milk you drank during the morning milk break.

V-7. Soda bottle measure

Purpose: To provide experience with milliliter measures

Materials: Empty soda bottle and graduated beaker or cylinder

Number of participants: Any number

Directions: Fill an empty soda bottle with water and pour it into a graduated container to determine how many milliliters of soda were in the bottle.

V-8. Preparation of 1-liter container

Purpose: To provide experience with 1-liter amounts

Materials: Known liter container, quart milk carton, scissors, and pencil

Number of participants: Any number of teams of 2 children

Directions: Measure 1 liter of water in the known liter container. Pour this water in a quart milk container that has the top unfolded. Cut off the unused portion of the container before marking the water level on the inside with a pencil. (Since the container expands when the top is removed, the mark must be made after the top is removed.) The resulting container may be used as a known 1-liter container.

V-9. Preparation of a 2-liter container

Purpose: To provide experience in preparing a 2-liter container

Materials: Known liter container, one-half gallon milk container, scissors, and pencil

Number of participants: Any number of teams of 2 children

Directions: Pour 2 liters of water using the known liter measure into the one-half gallon milk carton. Cut off the unused portion of the top of the carton. Mark the water level on the inside with a pencil. The resulting container may be used as a known 2-liter container.

V-10. Preparation of a 4-liter container

Purpose: To provide actual experience in preparing a 4-liter container

Materials: Known liter container, 1-gallon milk carton, scissors, and pencil

Number of participants: Any number of teams of 2 children

Directions: Pour 4 liters of water from the known liter measure into the 1-gallon milk carton.

Cut off the unused portion of the top of the carton. Mark the water level on the inside with a pencil. The resulting container may be used as a known 4-liter container.

V-11. Build a solid

Purpose: To acquaint children with the concept of metric volume and how small building blocks make a solid

Materials: Playing board from A-6, Cuisenaire rods or metric blocks, and a die

Number of participants: 2 to 4

Directions: Play begins by rolling the die to determine what metric volume, that is, 1 to 6, may be placed on the playing board. The winner is the first person to exactly complete the 1 cm by 10 cm by 10 cm solid. No piece may be moved after it is played and the player must take the exact amount rolled each time in one piece.

V-12. Build a cube

Purpose: To provide the concept of metric volume by showing how a 3-centimeter cube is built of many smaller cubes

Materials: One box of Cuisenaire rods or other metric blocks and a die with sides taped and labeled so that 1, 2, and 3 appear twice

Number of participants: Groups of 2 to 4

Directions: Place the white, red, and light green Cuisenaire rods in front of the players. The players roll for high score to determine who will start. The object of the game is to see who can first build an exact 3-centimeter cube. The spots on the die determine the length of the rod to be played. If a player rolls a 3, he may take a light green, red and white, or 3 whites. After the rods are once played, they cannot be moved.

V-13. Volume relay

Purpose: To provide advanced activities in stating equivalent units of metric volume

Materials: None

Number of participants: Total class divided into two teams

Directions: One member of each team goes to the chalkboard as the game leader places on the board the amount such as 200 ml. One point is scored for each correct equivalent unit written before the leader can count slowly to 10. (Correct answers would be 20 cl and 2 dl.)

V-14. Super cross metric puzzle

Purpose: To provide abstract activities in adding metric volume

Materials: Accompanying puzzle

Number of participants: Any number

Directions: To complete "a" down, determine how many liters must be added to 1 kiloliter to equal 2 kiloliters. Continue until all blanks are filled. (Have students make additional puzzles.)

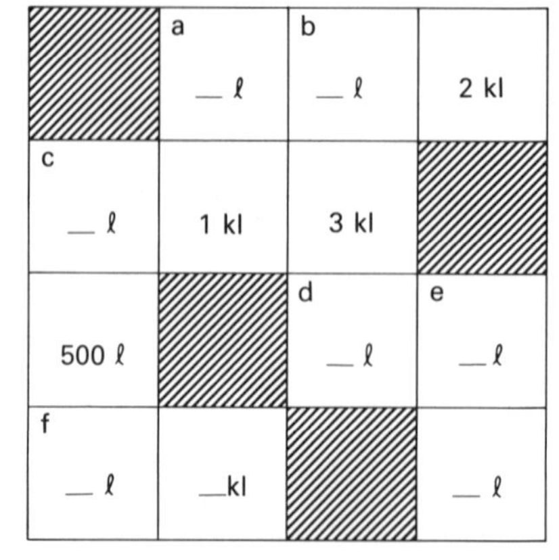

Down: add to
a. 2 kl
b. 5 kl
c. 2 kl
e. 1.75 kl

Across: add to
a. 4 kl
c. 4.5 kl
d. 1.5 kl
f. 3.5 kl

V-15. Blow a liter of hot air—always a favorite

Purpose: To make a cubic decimeter that holds 1 liter

Materials: Sheets of newsprint that form a square 40 centimeters on a side

Number of participants: Any number

Directions:

1. Fold the paper from corner to corner as in Diagram 1.
2. Fold the sides in to form a triangle as shown in Diagram 2.
3. Fold points E and B down to point A. Turn it over and fold C and D down to point A. Keep point A pointing downward as in Diagram 3.
4. Corners H and F are now double and A is loose. Fold H and F to meet in the center. Turn it over and do the same for the corners on the back, Diagram 4.
5. Fold the loose ends out to form Diagram 5. Turn it over and do the same on the back.
6. Fold J and K over to the middle to form right triangles. Turn it over and do the same for the back, Diagram 6.
7. Tuck the small right triangles made in number 6 into the pockets along LM and LN. Do not tuck them under LM and LN. Turn the model over and do the same on the back.
8. Open the hole at G slightly with a pencil point. Blow hard into this hole, forming and creasing the cube as you blow. The finished cube will be 10 centimeters on a side and will hold 1 liter of air.

1

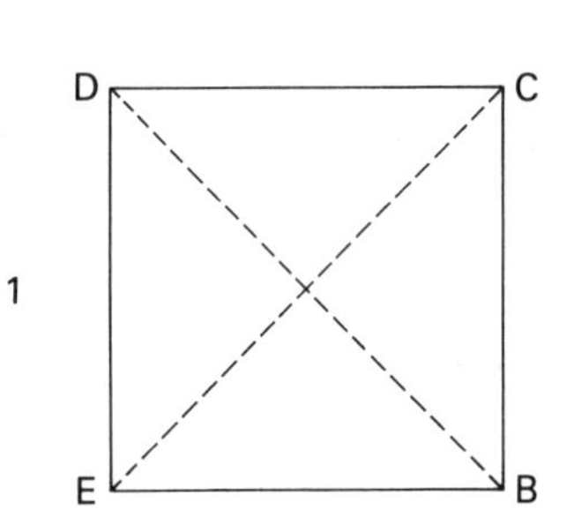

2

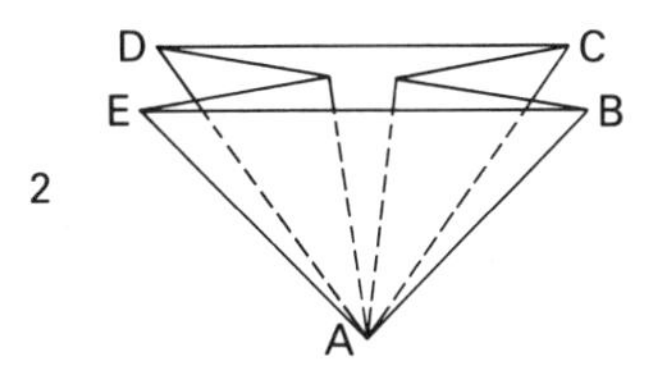

3

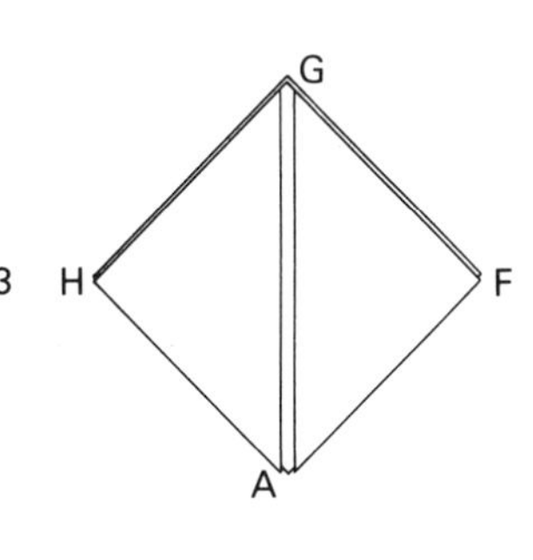

4

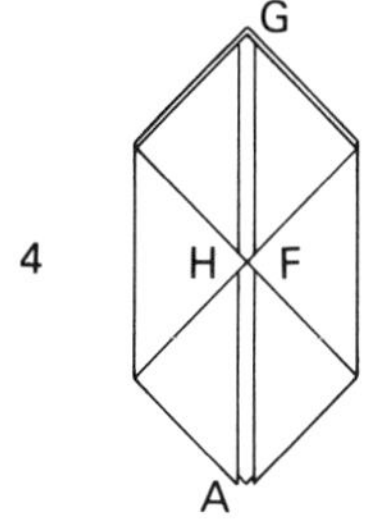

5

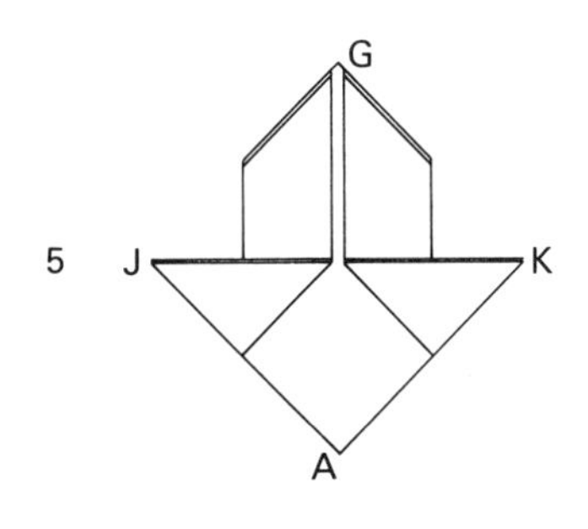

6

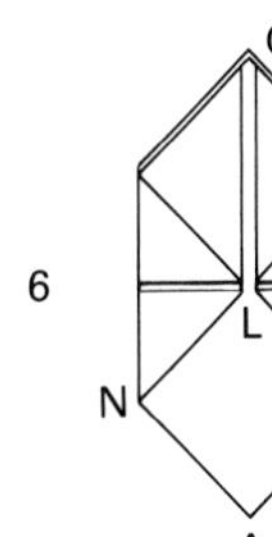

V-16. Spin a kiloliter

Purpose: To provide advanced activities in adding units of metric volume

Materials: Spinner made from a pizza board or the top of a cottage cheese container as follows:

Number of participants: 2 to 4

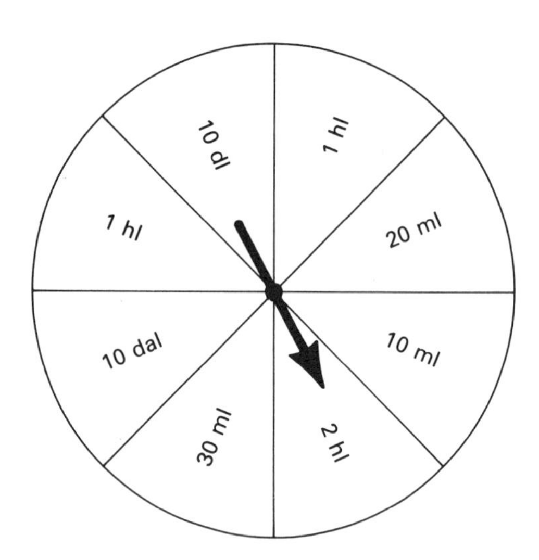

Directions: Each player spins 2 times and adds the 2 volumes. Play continues until someone reaches a total score of 1 kiloliter. (Remember a cubic meter would contain a kiloliter of air.)

V-17. Abstract conversion

Purpose: To provide experience converting within the metric system

Materials: Accompanying problems

Number of participants: Any number

Directions: Convert the following as indicated.

a. 1 ℓ =	________ ml	i. 24.3 ml =	________ cl
b. 1 cl =	________ ml	j. 25 dal =	________ hl
c. 1 kl =	________ ℓ	k. 0.0134 kl =	________ ml
d. 12 hl =	________ ℓ	l. 0.142 dal =	________ cl
e. 1.3 cl =	________ dal	m. 143.62 dal =	________ dl
f. 0.001 kl =	________ dal	n. 243.6 ℓ =	________ ml
g. 0.03 hl =	________ cl	o. 69.2 kl =	________ dl
h. 0.01 ℓ =	________ cl	p. 72.3 kl =	________ dal

Answers:

a. 1,000	e. 0.0013	i. 0.0243	m. 14,362
b. 10	f. 0.1	j. 2.5	n. 243,600
c. 1,000	g. 300	k. 13,400	o. 692,000
d. 1,200	h. 1	l. 142	p. 7,230

CHAPTER 5

TEACHING TIME AND TEMPERATURE

TIME

The international system of units (SI) includes the second as one of the base units of the metric system. For this reason a section on teaching time is included to give suggestions and games for aiding children in learning to tell time. Actually the adoption of the metric system will add nothing new in the measurement of time as the units will remain identical, that is, second, minute, hour, and day. However, teaching the measurement of time has always involved difficulties, as the ability to tell time requires a maturation that is not reached at the same age in every child. Thus teaching this subject spans several years and can be a threatening experience to children.

The teaching of time measurement is usually begun in first grade where varying degrees of success are achieved. The second-grade teacher starts over in the time-telling process but is able to progress a little farther than the teacher of the previous year. The third-grade teacher also finds it necessary to give considerable emphasis to telling time. At this level she finds considerable variation in the children's ability to tell time. Some did learn to tell time in the previous grades, some may have received a watch as a bribe for learning to tell time with help from brothers, sisters, and grandparents, and some still do not know how and feel threatened.

Jean Piaget[1] has found in his studies that children of 5 to 9 years of age are at a loss in working with watches and sandglasses, since the children believe that the motions vary with the actions being timed. This concept is exemplified by a 6-year-old child who cannot conserve time. He believes that it takes more time for the sand to run out of a sandglass if a task such as moving marbles from one vessel to another is done slowly. Children in this stage need readiness activities for telling time rather than the frustration of actually attempting to tell time. Copeland[2] suggests some easily administered tasks to determine who can conserve time and who cannot. This knowledge would aid in meeting the

[1]Jean Piaget: The child's conception of time, New York, 1969, Basic Books, Inc., p. 176.
[2]Richard Copeland: How children learn mathematics, New York, 1974, MacMillan Publishing Co., Inc., pp. 175-182.

specific needs of children and prevent undesirable results caused by providing an abstract lesson to children who cannot conceive of time.

Readiness activities

There is a tremendous temptation for teachers to move directly to the specific objective of telling time on the hour, on the half-hour, and then on the quarter-hour. This objective should be postponed while readiness activities are introduced to help children conceptualize specific amounts of time. To help children understand the length of an hour, the teacher could set an alarm clock to go off each hour for an entire day. If a timer is available, the time intervals could be reduced to a few minutes. During these activities, considerable discussion should take place using the specific terms such as hour, minute, and 5 minutes. The timer should be taken out at recess and set for the proper number of minutes. It could be set for a 10-minute story period. To give children a feel for estimating time, ask the class members to lay their heads on their desks and without looking at others, raise their heads when they feel 1 minute is up. To improve the children's ability to estimate a minute, give the instructions to try the experiment again, but this time count to 60 before raising their heads. The results from this latter method should much more closely resemble the duration of 1 minute.

The following activities are designed to provide children the opportunity for active involvement in the process of gaining the concept of time.

TIME ACTIVITIES

Ti-1. Human arm clock

Purpose: To introduce children to the large hands and small hands of a clock by using a kinesthetic approach

Materials: Demonstration clock

Number of participants: Small group or whole class

Directions: Ask the children to stand by their seats and follow you in the demonstrations. Explain that the short hand will be made by bending the arm at the elbow and the long hand will be fully extended. After this is practiced, the children may wish to make the times of 9, 12, and 3 o'clock, since these are easily made with human arms.

Ti-2. Lay-on-the-floor clock

Purpose: To let children actively participate in forming the hands of a clock

Materials: Clock face drawn on a large piece of cardboard such as an unfolded refrigerator shipping carton. (Or face may be represented on the floor with tape or chalk.)

Number of participants: Whole class

Directions: Divide the class into long (minute) hands and short (hour) hands by actual height of children. Demonstrate how 2 of the children could lie on the clock to form the hands. Continue until each child gets to form the hands of the clock. On-the-hour times may be formed, such as 12, 3, 6, and 9 o'clock.

Ti-3. Floor-clock relay

Purpose: For children to become actively engaged in telling time

Materials: Two clock faces as in Ti-2. Cards with a time stated on the hour, such as 10 o'clock.

Number of participants: Whole class divided into 2 teams

Directions: After dividing the class into 2 teams, have each team divide equally into pairs of hour hands and minute hands. Place enough cards face down so that each pair can select a card, race to the clock, and form the time. The teacher and any children left over may act as referees for the game. The first team to finish the cards, wins the game. (A somewhat less active version of this game would direct the children to stand on the numerals to form the time as stated on the card.)

Ti-4. Make a clock face

Purpose: To gain the relative positions of the elements of a clock face

Materials: Paper plates, brads, dark-colored crayon, and tagboard for hands

Number of participants: Any number

Directions: While demonstrating an example of the paper-plate clock face, explain to the children that the 12, 3, 6, and 9 should be placed on the clock first, followed with the other numerals. The hands are secured with a brad. If it is desired that the clock faces be accurate, considerable guidance and assistance will be necessary. Sources of additional help could be mothers or upper-grade children.

The clock faces can be used by the children to duplicate a time that the teacher places on a clock face. More advanced activities could be to represent on the clock face a verbal time given by the teacher, such as 7 o'clock.

Ti-5. Tell the time

Purpose: To gain experience in telling time in everyday situations

Materials: Demonstration clock

Number of participants: Small groups or whole class

Directions: The teacher tells a story using the names of children in the class. When a child's name is used, he goes to the clock and records the proper time, using the movable hands. The beginning of such a story might be: Bryan awoke at 7 o'clock and was very hungry. However, since breakfast wasn't ready, he played with his little dog, Clipper, until 7:15. His mother, Mary, did not have breakfast ready until 7:30. Father, Bill, left for work at 8 o'clock. Bryan met the school bus at 8:30 for his ride to school.

Ti-6. Show the time

Purpose: To help children understand time

Materials: Old newspapers, magazines, and catalogs

Number of participants: Divide the class into teams of 4

Directions: Assign each team to make a collage of pictures showing what normally happens at certain hours each day, such as 12 o'clock noon, 7 o'clock in the morning, 4 o'clock in the afternoon, and 9 o'clock at night.

Ti-7. Do you know the time?

Purpose: To stress the practical, everyday uses of telling time

Materials: Demonstration clock with movable hands and old magazines, newspapers, and catalogs

Number of participants: Whole class

Directions: During phase 1 of this activity, the teacher should hold a class discussion concerning the answers to the following questions:

a. What time do you get up in the morning?
b. What time does school start?
c. What time do we go to lunch?
d. What time is school out?
e. What time do you go to bed?

During phase 2 the children would be given magazines, newspapers, and catalogs in which to search for pictures that show people getting up, going to school, eating, etc. These should be pasted in a "Telling Time" booklet. Appropriate description and times should be written below the pictures such as "Getting up in the morning, 7:30."

Ti-8. Clock bingo

Purpose: To reinforce time-telling skills with a game-like activity

Materials: For each child a bingo sheet that has 9 clock faces in a 3-by-3 array and markers such as small squares of construction paper. Since the clock faces do not have hands, all sheets are the same and therefore may be duplicated to save teacher time. Before starting the game, ask the children to draw on-the-hour hands on their clock faces. Stress that everyone's card should be different and that no time should appear more than once. If the children are not able to correctly complete the clock faces, invite some older children to assist with the activity.

Number of participants: 2 or more

Directions: There are various ways to play the game with the least mature being recognition of a time that the teacher puts on her demonstration clock. The first child to get 3 markers in a line wins the game.

Another version would be for the teacher to call out a time, such as 10 o'clock. The ways of winning may also be altered by stipulating that only the following way can win: (a) 3 in a horizontal line, (b) 3 in a vertical line, (c) 3 in a diagonal line, (d) all 4 corners, and (e) blackout. Small prizes may be given, with the rule that no one can win more than one prize. At the end of the playing period, the caller could continue to call until everyone wins.

Ti-9. Charades

Purpose: To provide meaning to specific hours of the day

Materials: None

Number of participants: Divide the class into teams of 4

Directions: The teacher will meet with each group and assign an hour that the other teams will try to guess as the activities of the hour are acted out. If the hour were 6 o'clock in the evening the first clue might show eating but no hint as to which meal would be given. Additional clues are given until someone can guess the hour. (A limit may be placed on the number of guesses.)

Ti-10. A.M. and P.M.

Purpose: To provide research opportunities for selected students

Materials: Encyclopedias

Number of participants: 1 or 2

Directions: Just prior to the lesson when the abbreviations A.M. and P.M. are to be introduced, select one or two students to give a report on these abbreviations. This information may be found in an encyclopedia under the entry "Time." (A.M. is for *ante meridiem,* before midday, and P.M. is for *post meridiem,* after midday.)

Other excellent topics for special reports are as follows:

a. United States time zones
b. Worldwide time zones
c. International date line
d. 24-hour clocks

Ti-11. Clock rummy

Purpose: To reinforce time sequence in a game-like activity

Materials: Ink pad and clock-face stamp to use in the preparation of cards for the game. If a stamp is unavailable, faces may be duplicated and then pasted on index cards. In preparing the cards, stamp one side of each of 48 index cards with a clock face. Pass out 2 cards to each of 24 children. Give specific directions such as, "You make 12:00 on your two cards," or "You make 12:30 on your two cards." In the end there should be 2 cards for each hour and each half hour.

Number of participants: 2 to 4

Directions: Begin the play by dealing 5 cards to each player and then placing one card face up to be used as the discard pile. Place the remaining cards face down. The player to the left of the dealer plays first. He may take the face-up card or one from the face-down pile. After drawing, he must discard unless he has played all his cards. The object is to complete a run of 3 such as 1:00, 1:30, and 2:00. Play continues until someone runs out of cards or until all cards in the face-down pile have been taken. Each run of 3 is worth 10 points. The winner must have 100 points. This will involve playing several hands.

TEMPERATURE

It is going to be another scorcher today with the temperature forecast to reach 37 degrees. Which will it be, Fahrenheit, centigrade, Celsius, or Kelvin? The statement and the following question indicate that the measurement of temperature holds potential confusion for both the younger and older learners. Even though the SI basic unit of temperature is the Kelvin degree, the Celsius scale is commonly used by those countries that use the metric system of weights and measures. The Kelvin scale is used in some scientific work, whereas the Celsius scale is used in everyday life in those countries. The comparison of the temperature scales in the accompanying diagram shows that the unit Celsius degree is equal to the unit Kelvin degree or in other words a 1-degree change in temperature on the Celsius scale is equivalent to a 1-degree change on the Kelvin scale. The main difference between the two scales is that the Kelvin scale has an absolute zero. This is the lowest point that temperature can go. It is defined as the temperature at which all molecular and atomic motion stops.

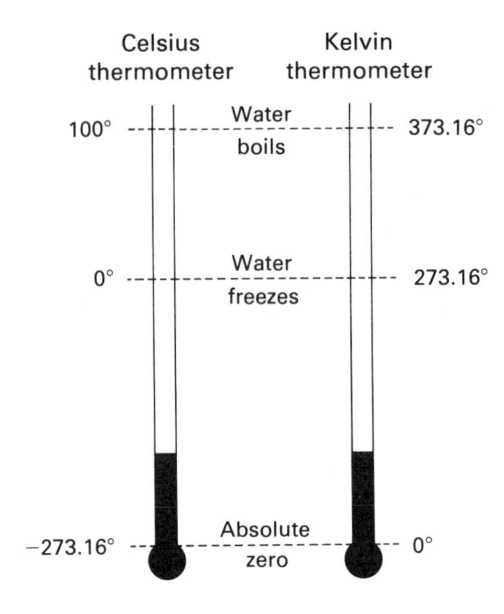

Changing terminology from the more familiar centigrade degrees to the less familiar Celsius degrees will result in confusion in the adult population. The change was done in deference to Anders Celsius, a Swedish astronomer who was the first to describe the centigrade or Celsius thermometer. The term "centigrade" comes from the fact that there are 100 equal units between the freezing point and the boiling point of water. If a student is looking for a reason for the change from the Fahrenheit to Celsius, a quick comparison of the gradations of the two scales may be all that is necessary. The 100 degrees between freezing and boiling on the Celsius scale are more easily handled than the 180 degrees on the Fahrenheit scale. Additionally, the Celsius scale is used throughout the world. Celsius degrees (° C = degrees Celsius) will be used in this chapter since this is the scale that will be used predominantly in this country as the metric system gains acceptance.

One characteristic of the Celsius scale, which can be considered a disadvantage, is that negative temperatures occur more often than on the customary Fahrenheit scale, such as a reading of 21° F would be −6° C. If children have not had instruction concerning negative or directed numbers, temperatures below zero can cause confusion. Some elementary science programs introduce temperatures below 0° C before children have had experience with the set of integers. As a readiness activity, children should be introduced to the thermometer as being a standing number line. It should be stressed that number lines need not be horizontal. This method of introducing negative numbers is a natural approach as the children can be shown that numbers below zero are negative and numbers above zero are positive. The two concepts negative numbers and temperature are so closely related that they should not be taught separately. Some recent attempts to integrate mathematics and science have stressed this type of "welding" together of related concepts.

The activities that follow are designed to alleviate the confusion that will arise concerning problems such as whether an outside temperature of 25° C would be conducive to a pleasant day on the beach. For a student to gain such an understanding, he must have varied opportunities to experience temperature.

TEMPERATURE ACTIVITIES

Te-1. Record the temperature

Purpose: To gain experience in taking actual temperature readings

Materials: Outdoor thermometer placed by a window in the shade so that a reading may be taken from inside the classroom

Number of participants: Teams of 2 children

Directions: Prepare a roster for 1 week for a team of 2 children to read the outside temperature each morning at 9 o'clock, and another team to read the outside temperature in the shade at 12 noon. This information should be recorded for seasonal and daily comparison. A graph of a week might look like the following:

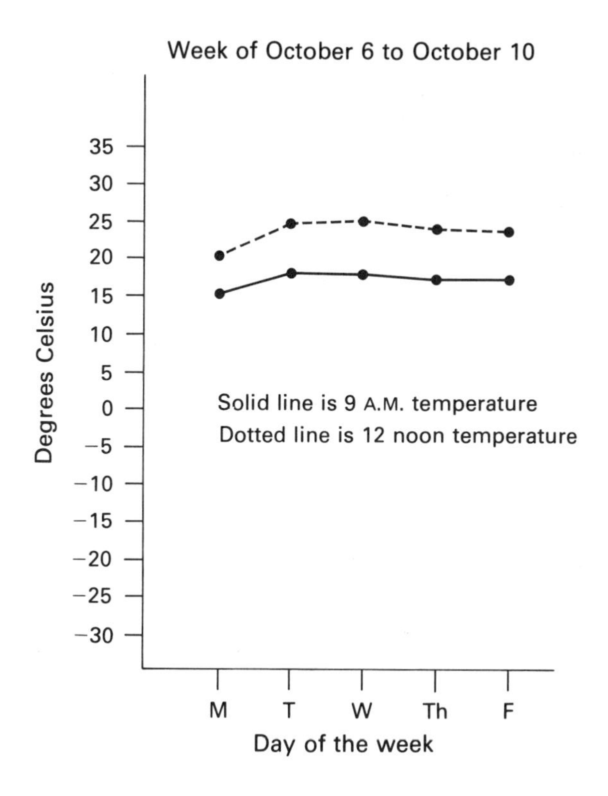

Te-2. Guess the temperature

Purpose: To promote awareness of the Celsius temperature scale

Materials: Celsius thermometers

Number of participants: Total class divided into teams of 2

Directions: After the children have had considerable experience reading temperatures both inside and outside, tell them that you want to have a contest to see who can come the closest in estimating the temperature. Provide each with a small piece of paper on which they are to record their estimate of the outside and inside temperatures. On this contest day the thermometers must be covered so that no one can read the temperature directly. (This contest should be repeated several times to reinforce the ability to estimate temperatures.)

Te-3. Your finger as a thermometer

Purpose: To provide children an opportunity to feel water of different temperatures

Materials: Dipping thermometer calibrated in degrees Celsius, 3 small vessels, ice, tap water, and hot water

Number of participants: Total class divided into teams of 2

Directions: Place at three numbered stations the following vessels of water: (a) a mixture of ice and water, (b) tap water, and (c) hot water. Ask the teams to go from station to station, and after they stick their fingers in the vessel, record their estimate of its temperature. The teacher or a designated team should measure the actual temperature of the water with the dipping thermometer and then report these findings to the class after all estimates have been made.

Te-4. Temperature of snow

Purpose: To demonstrate the temperature of snow

Materials: Snow, Celsius thermometer, and vessel (Freezer snow may be used by those in the southern areas of the United States.)

Number of participants: Demonstration to total class

Directions: If there is enough snow available, collect a vessel of snow and insert the thermometer to determine its temperature. Is there any relationship between the outside temperature and that of the snow?

Te-5. February contest

Purpose: To provide interest in temperature reading and recording

Materials: Outside thermometer calibrated in Celsius degrees

Number of participants: Total class

Directions: During the latter part of January, announce that there will be a contest. Explain that the class will be divided into 2 teams, the snow birds and the penguins. The school days for February will be divided equally on an alternating basis so that each day will belong to one one team or the other. The team having the day when the temperature at 9 o'clock is the lowest for the month will be the winner. The losing team is to provide a small party for the winners. If there is a tie, the teacher will put on the small party for the class.

Te-6. Things to do

Purpose: To create an understanding of Celsius temperature

Materials: Scissors, old newspapers, magazines, catalogs, large sheets of paper, and glue

Number of participants: Total class

Directions: Tell the class that they are to choose a temperature on the Celsius scale and prepare a collage of pictures depicting what you would do at that temperature. The heading might be "Things to do at 25° C."

Te-7. Temperature in Celsius

Purpose: To promote community awareness of Celsius temperatures

Materials: None

Number of participants: Total class

Directions: After the class has had considerable direct experience in using the Celsius temperature scale, they could make a request of a local radio or television weatherman to give the temperature in Celsius as well as Fahrenheit.

Te-8. Colder than 0° Celsius

Purpose: To demonstrate during a classroom experiment a thermometer reading below 0° Celsius

Materials: Crushed ice, tap water, salt, vessel, and a dipping thermometer

Number of participants: Demonstration to total class

Directions: Place a half-and-half mixture of crushed ice and water in a vessel. Place a dipping thermometer in the solution and record its temperature. After this has been done add a small amount of salt to the solution. After waiting a few minutes, make another temperature recording. The results should show that the temperature went below 0° C, indicating a clear need for negative numbers. Explain to the children that ice cream will not freeze unless salt is added to the ice and water mixture in the freezer.

Te-9. Telling the truth in temperature (Children love this activity.)

Purpose: To provide a class activity that promotes temperature awareness

Materials: Three large cards on which are recorded "I am zero degrees Celsius." The back of each card contains one of the following: "I am able to freeze water," "I am comfortable beach temperature," or "I am able to boil water."

Number of participants: 3 contestants, a panel of 4, with the remaining class members acting as an audience

Directions: The 3 contestants sit in front of the panel and each says, "I am zero degrees Celsius." Each contestant then gives one more clue such as "I am able to freeze water," "I am comfortable beach temperature," or "I am able to boil water." The panel then records the number of the person they think is the "real zero degree Celsius" on a pad and holds it up for the audience to see. The moderator then says, "Will the real zero degrees Celsius please stand." The game may be continued by using "I am 100 degrees Celsius." The clues can be body temperature, water boils, and very cold. The game can be made more difficult by using "I am the present outside temperature." To continue the game, new contestants and panel members may be chosen from the audience.

Te-10. Celsius diagnosis

Purpose: To diagnose children's understanding of the Celsius scale

Materials: None

Number of participants: Total class

Directions: At the end of the unit on temperature, ask the following questions:

a. If your temperature is 40° C, are you sick?
b. If the thermometer reads 25° C, will you need a short-sleeved shirt or a winter coat?
c. If the temperature is 30° C, should you go swimming or ice skating?
d. If the thermometer reads 0° C, will you need a winter coat?
e. If the thermostat at home is set at 20° C in winter, are you doing your share to conserve energy?
f. If the hot water is 20° C, will you have a hot, warm, or chilly bath?

g. If the temperature of a cup of hot chocolate is 90° C, will it burn your tongue?

h. Would it be likely to snow if the temperature were 15° C?

i. If the temperature reaches 40° C tomorrow will it be cool, warm, or hot?

Answers:

a. Yes
b. Shirt
c. Swimming
d. Yes
e. Yes
f. Chilly
g. Yes
h. No
i. Hot

CHAPTER 6

LEFTOVERS

This final chapter contains metric material that did not fit into the organization of the previous chapters. The games and activities presented in this chapter cover the general topic of the metric system rather than a specific metric topic such as length. These general activities are best used as a review when the children have reached a reasonable degree of metric awareness. Some of the activities may be used as an informal pretest to determine the extent of metric concepts already acquired by the children.

Many of the activities presented in this chapter are of a symbolic nature and do not rely on the use of concrete materials. Since these activities are more advanced, it is recommended that children first become involved with the concrete measuring activities presented in previous chapters before completing the general activities in this chapter.

The bulletin-board ideas are designed to motivate children to want to learn about the metric system. A metric bulletin board should be put up several days preceding the class work on the metric system. When children start asking questions about the terms, it is time to start the metric unit.

The last section of the chapter contains metric recipes. The coconut treats may be prepared without the use of a heat source. Some of the recipes require an electric skillet, whereas others require more sophisticated equipment. These recipes were developed and tested by a third-grade boy and his mother. He alone prepared the pancakes at home and reported that this was his favorite recipe.

LEFTOVER ACTIVITIES

O-1. Tic-Tac-Toe

Purpose: To provide children with an opportunity to use the various metric abbreviations

Number of participants: 2

Materials: 18 small squares of construction paper (9 red and 9 blue) with metric abbreviations of length, weight, and volume written on one side. Needed would be 3 red and 3 blue squares with metric abbreviations cm, m, and km; 3 red and 3 blue with mg, g, and kg; and 3 red and 3 blue with ml, ℓ, and kl. Also needed would be 3 direction cards to lay beside the playing area on which are printed length, weight, and volume.

Directions: The game follows rules similar to tic-tac-toe with a player choosing either red or blue squares to place on the playing board. The abbreviation to be placed on the board must correspond to the horizontal unit, such as length. As a player plays an abbreviation, he says the name; for example, for "ml" he would say "milliliter." The first player to get 3 of his color in a line, wins the game.

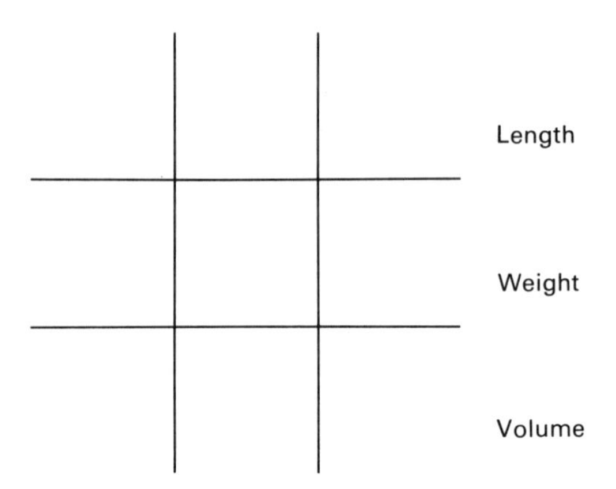

O-2. Metric Kool-Aid

Purpose: To provide practical experience in working with units of metric weight and volume

Materials: Kool-Aid, sugar, water, balance scales, and milliliter measure

Number of participants: Any number of students working in pairs

Directions: Prepare the following recipe to obtain one glass of drink:

1 g Kool-Aid
33 g sugar
250 ml water

O-3. Rain gauge

Purpose: To provide experience in working with metric measurements

Number of participants: One committee of approximately 4

Materials: Test tube, masking tape, and a wooden or wire holder

Directions: Make a rain gauge to place on the school grounds to measure the rainfall in centimeters and milliliters. Place a strip of masking tape vertically on the tube. Record the centimeter markings on one side of the tape and the milliliter markings on the other side. The committee can devise its own way of reporting the rainfall to the class and school. A weather bulletin board could be placed in a central location where the entire school could observe it.

O-4. Metric concentration

Purpose: To provide practice in recognizing metric symbols

Materials: 36 squares of heavy paper 5 cm on each side with the notations as shown in the example.

Number of participants: Any number

Directions: Play starts with all squares turned face down. Each player is permitted to ask to turn over 2 squares. If they are the same he may keep them and try 2 more. When he fails to get a match, the next player gets to choose. The wild cards will match with any symbol. The child with the most cards at the end wins.

m	cm	dg	ℓ	dag	kg
ml	dag	Wild card	dam	mm	cg
dm	mg	km	°C	hg	m
kg	Wild card	hm	mm	mg	km
cm	dg	ℓ	cg	Wild card	dm
Wild card	°C	hm	hg	dam	ml

O-5. Metric golf course

Purpose: To provide experience in following metric directions

Materials: Typing paper and metric rulers

Number of participants: Total class working individually

Directions: (Teacher reads as children make constructions.) Construct a rectangular golf course 18 cm × 24 cm. Mark an entrance with a point on the bottom 4 cm from the lower right corner with the sheet lengthwise in front of you. The clubhouse boarders 2 sides of the course and is in the lower right corner. It is a square building and occupies an area of 4 square centimeters. Draw the outline of the building. Locate, mark, and number the 9 holes of the golf course according to the following directions.

Hole

1. 2 cm above the upper left corner of the clubhouse
2. 3 cm in and 10 cm up from the lower right corner of the course
3. 1 cm down and 5 cm left of upper right corner of course
4. 4 cm below the center of the upper border
5. 10 cm straight left of hole 4
6. 3 cm in and 7 cm down from upper left corner of course

7. 1 cm in from center of left side
8. 3 cm up and 10 over from lower left corner of course
9. 3 cm up and 4 left of entrance

Compare your golf course with that of your neighbor.

O-6. Metric money

Purpose: To provide a thought-provoking situation concerning the use of metric terms to describe money

Materials: None

Number of participants: Groups of 4

Directions: The treasurer of the United States has asked our class to help in changing the names of our units of money to correspond to the metric system. He believes that there should be kilodollar bills, hectodollar bills, dekadollar bills, dollar bills, decidollar coins, centidollar coins, and millidollar coins. What do you think of this proposal? What would these amounts be equal to in today's money? Would it be necessary to have additional units to those proposed? If so what would you propose? Make a report to the class concerning your recommendations.

O-7. How would you measure it?

Purpose: To provide awareness of the various metric units

Materials: Scissors, glue, old catalogs or magazines, and sheets of paper

Number of participants: Total class

Directions: This activity should follow weighing and pouring. Each child should have 3 sheets of regular paper, one labeled liters, one grams, and one meters. The children are to cut pictures of various items from the catalogs and magazines and paste them on the sheet, which states how the item would be measured; for example, candy would be placed on the sheet labeled grams. (This page may also have the term "kilograms" recorded on it if the term has been introduced.)

O-8. Metricate the school

Purpose: A class activity that will provide metric experience for others in the school

Materials: Vary according to activity selected

Number of participants: Small groups or whole class

Directions: Suggest to the class that it would be nice if we could prepare materials that would help the whole school gain a knowledge of the metric system. Suggest that class projects could include the following:

a. Preparing a newspaper that might include problems and explanations of the metric system
b. Preparing for the library booklets that would give information about the metric system
c. Preparing cassette tapes that would tell the metric story
d. Preparing some metric card games to distribute to classrooms
e. Inviting someone who has lived in a country where the metric system is used to speak to classes
f. Obtaining an outstanding metric film that could be shown to the entire school

O-9. Farm problem

Purpose: To provide computational experience with metric units

Materials: Pencil and paper

Number of participants: Total class

Directions: Work the following problem:

If a farmer harvested 1.7 metric tons of oats per hectare from his 3.5 hectare field, how many metric tons did he harvest? (Make additional problems and place them on index cards.)

O-10. Cross-metric puzzle

Purpose: To provide a diagnosis of metric skills

Materials: Puzzle shown below

Number of participants: Any number

DOWN

1 The prefix used in the metric system meaning one tenth (1/10) is ______.

3 1/1,000 of a meter is equal to 1 ______.

4 1 meter is the same as 1,000 ______.

6 The abbreviation for 1/10 gram is ______.

7 The abbreviation for decimeter is ______.

8 The standard unit of weight in the metric system is the ______.

11 10 ______ s are equal to 1 meter.

12 Another way of saying 100 meters is 1 ______.

14 The abbreviation for 1/1,000 of a meter is ______.

16 The metric system is based upon the base ______.

19 The international way of spelling "meter" is ______.
21 The abbreviated form for kilometer is ______.
23 1,000,000 grams is equal to the metric ______.
24 The abbreviation for 100 meters is ______.
26 Centiliter can be abbreviated as ______.
28 Dekameter can be abbreviated as ______.
29 1/10 liter is abbreviated as ______.

ACROSS

2 1,000 meters is equal to 1 ______.
4 A kilogram weighs ______ than a gram (more, less).
5 In the metric system, time is based upon the solar ______.
9 A glass holding 1 centiliter of water is ______ than a glass holding a liter of water.
10 1,000 ______ s are equal to 1 gram.
12 The abbreviation for hectogram is ______.
13 The standard unit for length in the metric system is the ______.
15 Another way of saying 1 dekameter is 100 ______ s.
17 A unit of volume is the ______.
18 The abbreviated form for centimeter is ______.
20 1 kilometer is equal to a ______ meters.
22 Another way of saying 1/100 of a meter is ______.
25 The system of measurement based upon the decimal system is the ______ system.
27 1 gram is equal to a ______ centigrams.
30 The abbreviation for milliliter is ______.

Answers:

Down				Across			
1	deci	16	ten	2	kilometer	17	liter
3	millimeter	19	metre	4	more	18	cm
4	millimeters	21	km	5	second	20	thousand
6	dg	23	ton	9	less	22	centimeter
7	dm	24	hm	10	milligram	25	metric
8	gram	26	cl	12	hg	27	hundred
11	decimeter	28	dam	13	meter	30	ml
12	hectometer	29	dl	15	decimeter		
14	mm						

O-11. Metric wheels

Purpose: To provide reinforcement activities in the recognition of metric symbols

Materials: 3 wheels made from pizza cardboard plates, one each for length, weight, and volume. The length wheel is shown in the diagram. The other 2 wheels are made in a similar manner. If desired, each wheel may be painted with 8 different fluorescent colors. The metric terms are written over the colors with a permanent marker. Spring clothespins are marked with the following symbols, which will match with a term on the wheel.

Length	*Weight*	*Volume*
mm	mg	mg
cm	cg	cg
dm	dg	dg
m	g	g
dam	dag	dag
hm	hg	hg
km	kg	kg

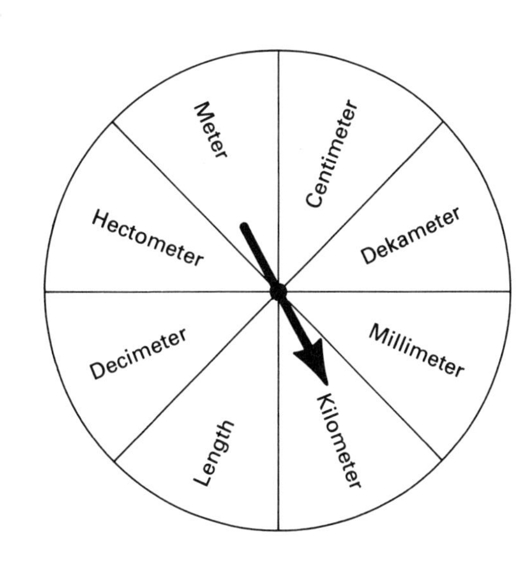

Number of participants: 3 relay teams of 7 members each

Directions: The wheels and pins are placed in the front of the room with the 3 relay teams lining up at the back of the room. The winner is the first team to correctly place the clothespins on the sections. (The activity could be made more difficult by placing length, weight, and volume terms on each wheel.)

O-12. Making metric words

Purpose: To familiarize students with metric terms

Materials: Accompanying words

Number of participants: Any number

Directions: Write a metric word on the space provided by taking the specified fractional part from the beginning of each word; for example 2/4 of "meat" would be "me."

2/4 meat
1/8 tomorrow
2/5 error ______(meter)______

1/4 decision
1/3 circus
1/2 metric
2/3 ere ____________

1/2 kind
1/3 lounge
2/2 me
1/2 tear
1/6 return ____________

3/5 gravy
1/4 meat ____________

1/2 kite
1/2 load
3/5 graph
1/5 motor ____________

5/7 million
3/5 grass
1/4 most ____________

3/7 century
2/7 tillage
1/2 little
2/5 error ____________

2/5 voice
2/5 lucky
1/3 metric ____________

2/5 light
1/4 tree
1/4 exit
1/3 rat ____________

.4 grass
.2 ammunition
.2 smash ____________

3/8 December
2/6 ignore
3/7 rambler ____________

1/2 mice
2/5 llama
2/3 ill
1/2 item
1/3 errand ____________

1/3 little
1/4 tree
1/3 eat
1/6 ration ____________________

1/3 decide
1/4 keep
1/3 always
2/2 it
1/3 errand

1/2 decent
2/3 ill
2/2 it
2/5 erase ____________________

3/4 kill
1/2 of
2/4 live
3/4 term ____________________

1/2 deep
1/2 kale
1/2 meat
3/4 term ____________________

1/2 help
1/3 can
2/3 too
4/7 grammar ____________________

3/4 milk
2/5 light
1/2 meal
3/8 terrible ____________________

O-13. Metric quiz

Purpose: To quickly diagnose metric competencies

Materials: Accompanying test

Number of participants: Any number

Directions: Choose the best answer for each of the following metric questions.

(1) The height of Mr. Brown, the principal, who is an average-sized man, would be
(a) 175 cm.
(b) 1 m.
(c) 240 cm.

(2) The weight of an M & M candy would be approximately
(a) 10 g.
(b) 2 dag.
(c) 1 g.

(3) The volume of a 1-liter milk carton would be
(a) 1,000 ml.
(b) 100 dal.
(c) 10 cl.

(4) A comfortable beach temperature would be
(a) 25° Celsius.
(b) 50° Celsius.
(c) 72° Celsius.

(5) If the service station attendant added a can of oil to your car, it would probably be
(a) 2 ml.
(b) 200 ml.
(c) 1 liter.

(6) The weather report indicates that tomorrow's high should be approximately 37° Celsius. Should you
(a) wear a winter jacket?
(b) wear a cool outfit?
(c) wear a light sweater?

(7) A tank that is 1 meter on each side would hold the following amount of water.
   (a) 1,000 kg
   (b) 1 metric ton
   (c) Both (a) and (b)
(8) The weight of an average-sized lady would be
   (a) 3,500 g.
   (b) 52 kg.
   (c) 25 kg.
(9) The temperature on the desert at midday could be
   (a) 40° Celsius.
   (b) 60° Celsius.
   (c) 90° Celsius.
(10) The playing area of a football field would contain approximately
   (a) 11 hectares.
   (b) 2 ares.
   (c) 45 ares.

Answers:

1 a	6 b
2 c	7 c
3 a	8 b
4 a	9 a
5 c	10 c

O-14. Unbaked cookies

Purpose: To provide an opportunity to use metric measures in cooking

Materials: Metric scales, milliliter measurer, saucepan, teaspoon, access to a stove, waxed paper, and ingredients as listed under directions

Number of participants: Individuals or small groups

Directions: Mix in a saucepan the following ingredients:

380 g of sugar	110 ml of milk
35 g of cocoa	113.4 g or one stick of margarine

Bring to a boil and cook 1 minute. Remove from heat and add:

240 g of oatmeal (quick)	5 ml of vanilla

Drop by spoon on waxed paper.

O-15. Rice Krispies treats

Purpose: To provide an opportunity to use metric measures in cooking

Materials: Metric scales, large saucepan, cake pan, access to a stove, and ingredients as listed in the directions

Directions:

58 g of margarine	135 g of Rice Krispies
200 g of miniature marshmallows	

Measure margarine into large saucepan and melt over low heat. Add marshmallows and cook. When marshmallows are melted, add Rice Krispies. Stir until well coated. Spread in pan and cut when cool.

O-16. Coconut treats

Purpose: To provide an opportunity to use metric measures in preparing a tasty treat

Materials: Metric scales, mixing bowl, spoon, knife, and ingredients listed under directions. No heat is required.

Number of participants: 6

Directions: Mix in bowl the following ingredients:

32 g of soft margarine
107 g of brown sugar
22 g of flaked coconut
6 square graham crackers

Cream margarine and sugar. Stir in coconut. Spread on graham crackers. Serves 6.

O-17. Fluffy popcorn

Purpose: To provide an in-class experience cooking with metric ingredients

Materials: Metric scales, electric skillet, ingredients listed under directions

Number of participants: 8

Directions: Heat 10 g of shortening or oil in an electric skillet set at medium high. Add 55 g of popcorn. Shake easily until popping has stopped. Salt to taste.

O-18. Peanut clusters

Purpose: To provide a metric experience in cooking

Materials: Metric scales, double boiler, plate, heat source, and ingredients listed under directions

Number of participants: 20 to 25

Directions:

100 g of semisweet chocolate chips
70 g of butterscotch chips
100 g of Spanish peanuts

Melt butterscotch and chocolate chips together in top of the double boiler. Stir in peanuts and pour onto a plate. When cool, break into pieces. Serves 20 to 25.

O-19. Pancakes

Purpose: To provide a cooking experience using metric measures; an activity designed for use at home

Materials: Metric scales, milliliter measurer, mixing bowl, heavy griddle, heat source, and ingredients listed under directions

Number of participants: 1 or more

Directions:

1 egg
275 ml of milk
13 g of baking powder
185 g of flour
4 g of sugar
28 g of soft shortening
3 g of salt

Beat egg well and add milk. Stir in baking powder, flour, sugar, shortening, and salt. Beat until smooth. Fry pancakes on a heavy griddle.

O-20. Fudgesicles

Purpose: To provide a family experience in metric food preparation; recipe to be prepared at home for use in a school metric party

Materials: Metric scales, milliliter measurer, heating and freezing sources, molds, and materials listed under directions

Number of participants: 1 or more

Directions:

133 g package of regular chocolate pudding
440 ml of milk
70 g of sugar
220 ml of milk

Cook chocolate pudding as directed, using 440 milliliters of milk. Remove from heat and stir in sugar and additional milk. Pour into molds or ice cube trays. Freeze.

METRIC BULLETIN BOARDS

You can't *wiggle* out of it—
Go METRIC!

Materials: Gauze or bandages can be used for the body. Colored construction paper can be used for the letters and head.

Let's crow about the METRIC SYSTEM

Materials: White construction paper can be used for the body. Crumpled tissue paper or real feathers may be used for the feathers. This bulletin board is good to use in the springtime when talking about farm animals.

Materials: Colored construction paper can be used for the head, scooter, and letters. Felt can be used for the turtle shell. The shell may be 3 dimensional.

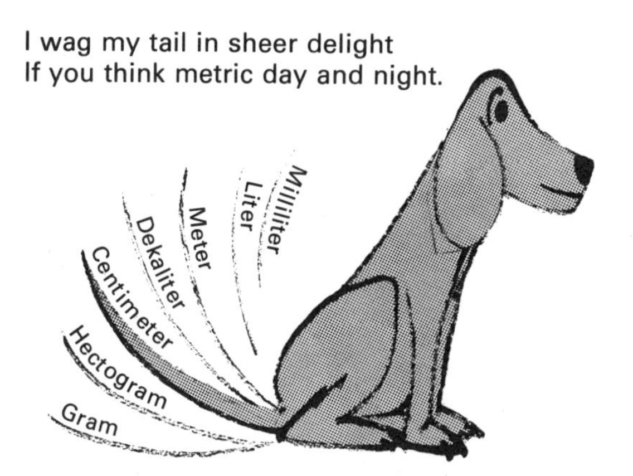

Materials: Felt or a furry material may be used for the dog. Pin the metric words on the dog's tail. Colored construction paper can be used for the letters.

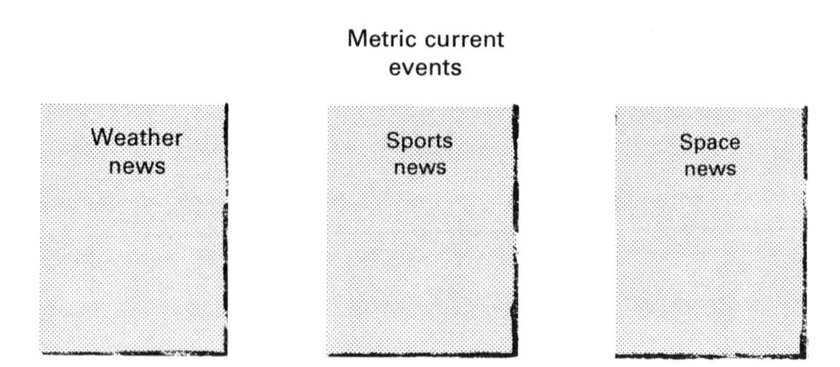

Materials: Letters can be made from newspaper. Have the class cut out clippings from magazines and newspapers that use measurement.

Metric around the house

Chocolate chips
340 g
Beans
340 g
Fabric softener
1.89 ℓ
Crunchy sugar puffs
340 g
Screws
50 mm
Cough syrup
120 ml
1 cm Seam allowance

Materials: Colored construction paper for the letters. Pin up labels from containers that have the contents recorded in metric units. (Ask the children to bring these to school and divide the class into teams for a contest.)

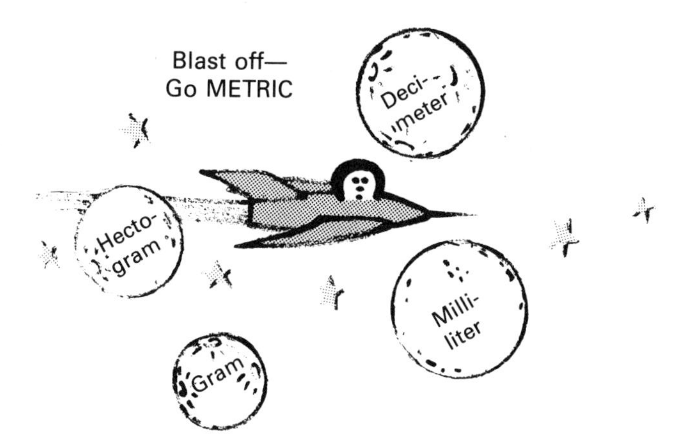

Materials: Dark blue paper can be used for the background. Tinfoil can be used for the stars. Colored construction paper can be used for the planets and letters.

PHRASES THAT WILL CHANGE

"Give him an inch and he'll take a mile.

"Traffic just inched along."

"Five feet two, eyes of blue."

"He hit the ball a mile."

"The quarterback threw a 20-yard pass."

"She bought a quart of milk."

"The baby weighed 7 pounds."

ARE YOU READY?

Meter

Materials: Newspapers can be used for the letters.

Materials: Felt can be used for the jack-in-the-box. Paper folded like an accordion and taped in the back of the jack-in-the-box can be used to make him spring out. Students who have mastered the metric system can have their name put on the bulletin board. Student's papers may also be put on the bulletin board. Colored construction paper can be used for the letters.

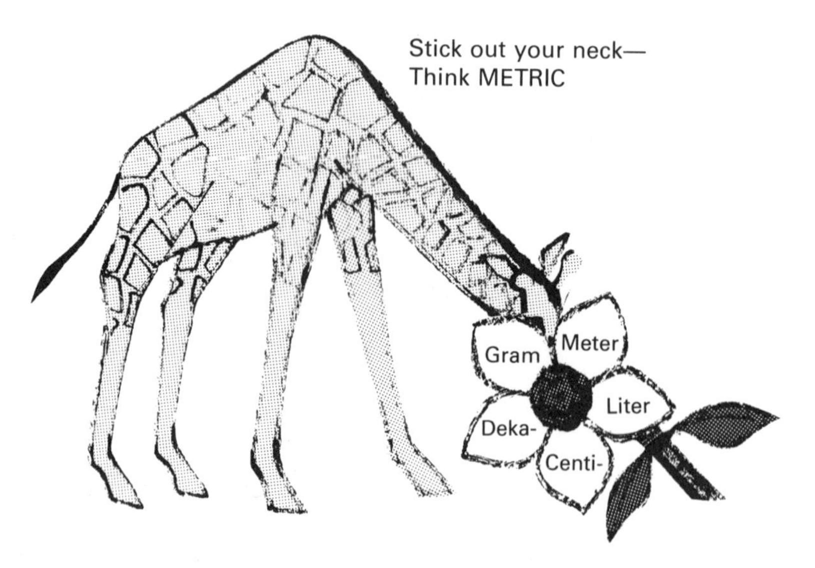

Materials: Gold-colored felt can be used for the giraffe. Tissue paper can be used for the flower. Paste the metric words on the flower. Colored construction paper can be used for the letters.

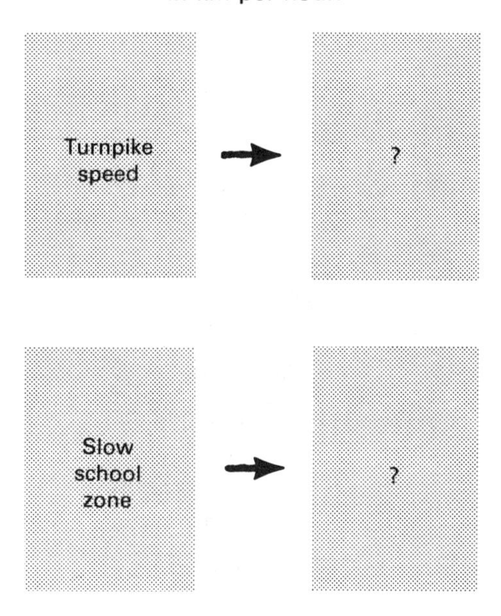

Materials: Poster board can be used for the traffic signs. Have the class decide what these signs should state.

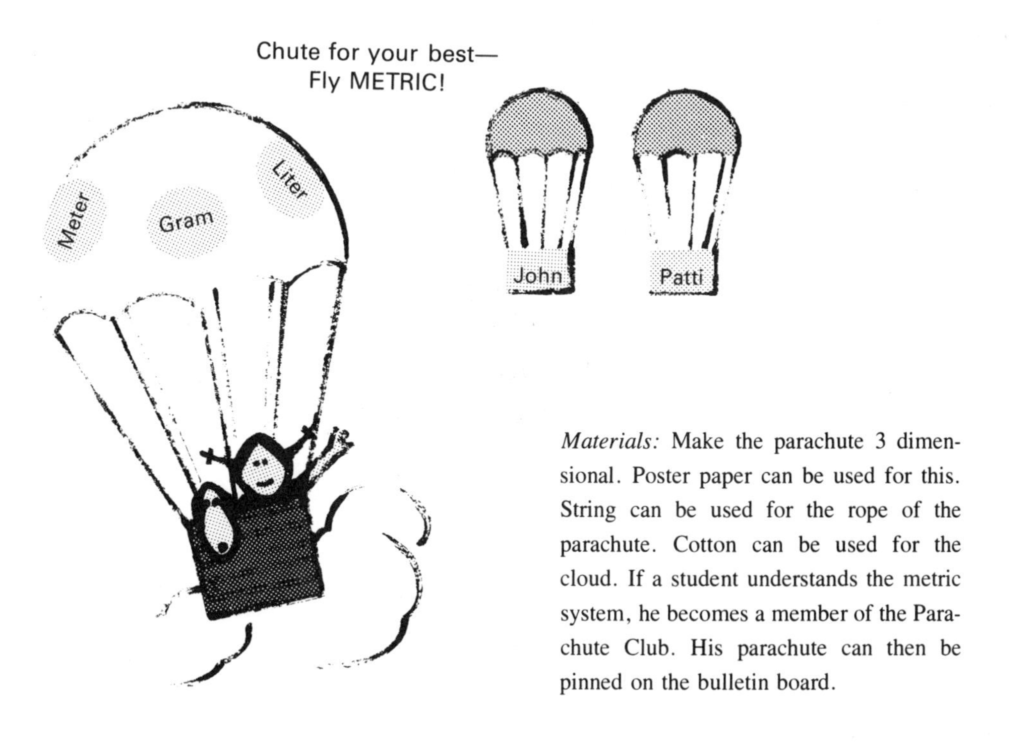

Materials: Make the parachute 3 dimensional. Poster paper can be used for this. String can be used for the rope of the parachute. Cotton can be used for the cloud. If a student understands the metric system, he becomes a member of the Parachute Club. His parachute can then be pinned on the bulletin board.

Materials: Use pipe cleaners for bug. Use colored construction paper for letters. Use cellophane on pipe cleaners for wings.

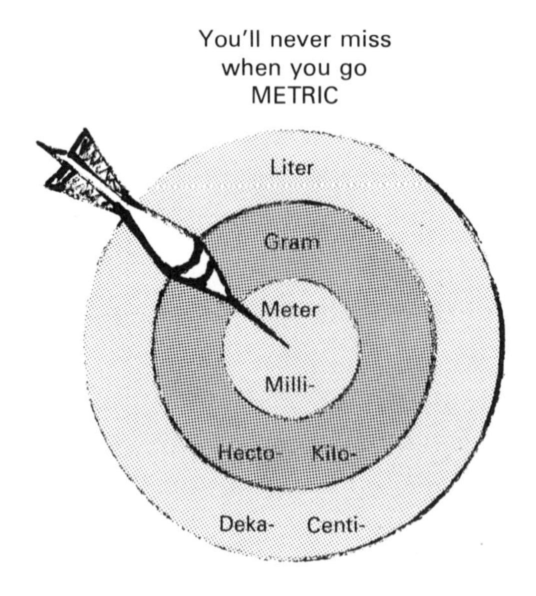

Materials: Use tinfoil for the dart board. Colored construction paper can be used for the letters. A real dart can be used for the arrow.

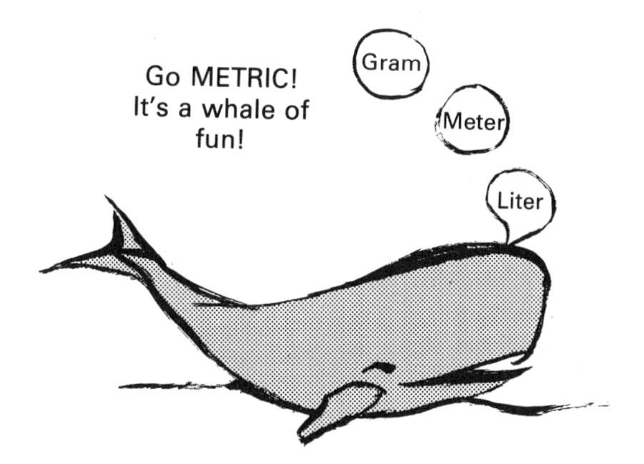

Materials: Red felt or construction paper for the whale. Blue tissue paper crumpled up for the water. Cellophane over the bubbles.

APPENDIX A

PRINCIPAL UNITS OF ENGLISH AND METRIC SYSTEMS

English system

LENGTH

12 inches = 1 foot
3 feet = 1 yard
16½ feet = 1 rod
5½ yards = 1 rod
5,280 feet = 1 mile
1,760 yards = 1 mile
320 rods = 1 mile

WEIGHT

437½ grains = 1 ounce
16 ounces = 1 pound
2,000 pounds = 1 ton
2,240 pounds = 1 long ton

AREA

144 square inches = 1 square foot
9 square feet = 1 square yard
160 square rods = 1 acre
640 acres = 1 square mile
1 square mile = 1 section

VOLUME

3 teaspoons = 1 tablespoon
16 tablespoons = 1 cup
2 cups = 1 pint
2 pints = 1 quart
4 quarts = 1 gallon
2 gallons = 1 peck
4 pecks = 1 bushel

Metric system

LENGTH

10 millimeters = 1 centimeter
1,000 millimeters = 1 meter
10 centimeters = 1 decimeter
100 centimeters = 1 meter
10 decimeters = 1 meter
10 meters = 1 dekameter
10 dekameters = 1 hectometer
10 hectometers = 1 kilometer
1,000 meters = 1 kilometer

WEIGHT

10 milligrams = 1 centigram
1,000 milligrams = 1 gram
10 centigrams = 1 decigram
10 decigrams = 1 gram
10 grams = 1 dekagram
1,000 grams = 1 kilogram
10 dekagrams = 1 hectogram
10 hectograms = 1 kilogram
1,000 kilograms = 1 metric ton

AREA

100 square millimeters = 1 square centimeter
100 square centimeters = 1 square decimeter
100 square decimeters = 1 square meter
100 square meters = 1 are
1 square dekameter = 1 are
100 ares = 1 hectare
100 hectares = 1 square kilometer

VOLUME

1 cubic centimeter = 1 milliliter
1 cubic decimeter = 1 liter
1,000 milliliters = 1 liter
1,000 liters = 1 cubic meter

METRIC UNITS OF LENGTH, WEIGHT, AND VOLUME

Prefix	*Linear measure*	*Weight*	*Volume*
$10^6 \times$ base	Megameter	Megagram	Megaliter
$10^3 \times$ base	Kilometer	Kilogram	Kiloliter
$10^2 \times$ base	Hectometer	Hectogram	Hectoliter
$10^1 \times$ base	Dekameter	Dekagram	Dekaliter
$10^0 \times$ base	Meter	Gram	Liter
$10^{-1} \times$ base	Decimeter	Decigram	Deciliter
$10^{-2} \times$ base	Centimeter	Centigram	Centiliter
$10^{-3} \times$ base	Millimeter	Milligram	Milliliter
$10^{-6} \times$ base	Micrometer	Microgram	Microliter

APPENDIX B

COMPARISON OF METRIC AND ENGLISH UNITS

LENGTH

1 millimeter = 0.03937 inches
1 centimeter = 0.3937 inches
1 centimeter = 0.0328 feet
1 meter = 39.37 inches
1 meter = 3.281 feet
1 meter = 1.094 yards
1 meter = 0.1988 rods
1 kilometer = 0.6214 miles
1 inch = 25.4 millimeters
1 inch = 2.54 centimeters
1 inch = 0.0254 meters
1 foot = 30.48 centimeters
1 foot = 0.3048 meters
1 yard = 0.914 meters
1 rod = 5.029 meters
1 mile = 1.6093 kilometers

WEIGHT

1 gram = 0.035 ounce
1 kilogram = 2.2 pounds
1 metric ton = 1.1 tons
1 ounce = 28.35 grams
1 pound = 453.59 grams
1 pound = 0.453 kilograms
1 ton = 0.907 metric tons

AREA

1 square inch = 6.45 square centimeters
1 square foot = 0.0929 square meter
1 square yard = 0.836 square meter
1 square mile = 2.59 square kilometers
1 acre = 0.4 hectare
1 square centimeter = 0.155 square inch
1 square meter = 1.2 square yards
1 square kilometer = 0.386 square mile
1 hectare = 2.47 acres

VOLUME

1 milliliter = 0.034 fluidounce
1 liter = 2.1 pints
1 liter = 1.06 quarts
1 liter = 0.264 gallon
1 liter = 0.028 bushel (dry)
1 fluidounce = 29.6 milliliters
1 teaspoon = 5 milliliters
1 tablespoon = 15 milliliters
1 pint = 0.47 liter
1 quart = 0.95 liter
1 gallon = 3.78 liters
1 bushel (dry) = 35.2 liters

APPENDIX C

CONVERSION TABLES, FORMULAS, AND RELATIONSHIPS

Conversion tables*

LENGTH

To change	*Multiply by*
miles to kilometers	1.6
miles to meters	1609.3
yards to meters	0.9
yards to centimeters	91.4
inches to centimeters	2.54
inches to millimeters	25.4
feet to centimeters	30.5
kilometers to miles	0.62
meters to yards	1.09
meters to inches	39.4
centimeters to inches	0.39
millimeters to inches	0.04

WEIGHT

To change	*Multiply by*
tons to metric tons	0.9
pounds to kilograms	0.45
pounds to grams	453.6
ounces to grams	28.4
metric tons to tons	1.1
kilograms to pounds	2.2
grams to pounds	0.002
grams to ounces	0.035

*The conversion factors have been rounded to aid in computations.

AREA

To change	*Multiply by*
square inches to square centimeters	6.45
square feet to square meters	0.093
square yards to square meters	0.836
acres to hectares	0.4
square miles to square kilometers	2.6
square centimeters to square inches	0.155
square meters to square feet	10.8
square meters to square yards	1.2
hectares to acres	2.5
square kilometers to square miles	0.386

VOLUME

To change	*Multiply by*
cubic inches to cubic centimeters	16.4
cubic feet to cubic meters	0.03
cubic yards to cubic meters	0.76
cubic centimeters to cubic inches	0.06
cubic meters to cubic feet	35
cubic meters to cubic yards	1.3
teaspoons to milliliters	5
tablespoons to milliliters	15
fluidounces to milliliters	29.6
pints to milliliters	473
pints to liters	0.47
quarts to liters	0.95
gallons to liters	3.8
bushels to liters (dry)	35.2
milliliters to fluid ounces	0.034
liters to pints	2.1
liters to quarts	1.06
liters to gallons	0.26
liters to bushels (dry)	0.03

Temperature conversion formulas

$F = (9/5 \times C) + 32$ or $F = (1.8 \times C) + 32$

To change F to C, subtract 32 from F and divide by 1.8.

$C = 5/9 \times (F - 32)$

To change C to F, multiply C by 1.8 and add 32.

Approximate relationships

1. The meter is a little longer than a yard.
2. The liter is a little more than a quart.
3. The kilogram is a little heavier than 2 pounds.
4. Body temperature is about 37° C.
5. A kilometer is about 0.6 of a mile.
6. A metric ton is a little more than a ton.
7. A hectare is about 2.5 acres.

APPENDIX D

METRIC SUPPLIERS, JOURNALS, AND NEWSLETTERS

METRIC SUPPLIERS*

A. Balla & Company, 3494 N. Ocean Boulevard, Fort Lauderdale, Florida 33308.

Activity Resource Company, Inc., P.O. Box 4875, Hayward, California 94545.

BFA Educational Media, 2211 Michigan Avenue, Santa Monica, California 94040.

Channing L. Bete Company, Incorporated, 45 Federal Street, Greenfield, Massachusetts 01301.

Creative Publications, P.O. Box 10328, Palo Alto, California 94303.

Cuisenaire Company of America, 12 Church Street, New Rochelle, New York 10805.

DCA Educational Products, 424 Valley Road, Warrington, Pennsylvania 18976.

Developmental Learning Materials, 7440 Natchez Avenue, Niles, Illinois 60648.

Dick Blick Company, P.O. Box 1267, Galesburg, Illinois 61401.

Edmund Scientific Company, Edscorp Building, Barrington, New Jersey 08007.

Educational Aids Department, Union Carbide Research Center, P.O. Box 363, Tuxedo, New York 10987.

Educational Teaching Aids Company, 657 Oak Grove Plaza, Menlo Park, California 94025.

Enrich, 3437 Alma Street, Palo Alto, California 94306.

Gallery Books, 1104 Lawrence Street, Los Angeles, California 90021. (For metric cookbook)

Goldstar, 3 Parkway Center, Suite 109 M, Pittsburgh, Pennsylvania 15220.

Harcourt Brace Jovanovich, Inc., Polk & Geary Streets, San Francisco, California 94109.

Hayes School Publishing Company, Incorporated, 321 Penwood Avenue, Wilkensburg, Pennsylvania 15221.

Idaho Research Foundation, Incorporated, Box 3367, University Station, Moscow, Idaho 83843.

Ideal School Supply Company, 11000 South Lavergne Avenue, Oak Lawn, Illinois 60453.

Imperial International Learning, Box 548, Kankakee, Illinois 60901.

Instructor Curriculum Materials, Instructor Park, Dansville, New York 14437.

Jay-Art Studios, Box 1520 M, G.P.O., Adelaide 5001, South Australia.

JEM Innovations, 4568 East 45th Street, Tulsa, Oklahoma 74135.

Laidlaw Brothers, Publishers, 701 Welch Road, Palo Alto, California.

La Pine Scientific Company, 6001 S. Knox Avenue, Chicago, Illinois 60629.

Larry Harkness Company, 115 North Princeton Avenue, Villa Park, Illinois 60181.

Lawhead Press, Incorporated, 900 E. State Street, Athens, Ohio 45701.

Learning Resource Center, Incorporated, 10655 S.W. Greenburg Road, Portland, Oregon 97223.

Library Filmstrip Center, 3033 Aloma, Wichita, Kansas 67211.

Listener Educational Enterprises, 6777 Hollywood Boulevard, Hollywood, California 90028.

MacLean-Hunter Learning Materials Company, 481 University Avenue, Toronto 101, Ontario M5W 1A7 Canada.

McGraw-Hill Ryerson Limited, 330 Progress Avenue, Scarborough, Ontario M1P 2Z5 Canada.

Media Materials, Incorporated, 409 West Cold Spring, Baltimore, Maryland 21210.

*Even though I have attempted to include all companies that sell metric aids, there obviously are omissions.

Metric Teaching Aids, 75 Horner Avenue, Toronto 530, Ontario, Canada.

Metrix Corporation, P.O. Box 19101, Orlando, Florida 32814.

Mind/Matter Corporation, P.O. Box 345, Danbury, Connecticut 06810.

Moyer-Vico Limited, 25 Milvan Drive, Weston, Ontario M9L 1Z1 Canada.

National Bureau of Standards, Metric Information Office, Washington, D.C. 20234.

National Council of Teachers of Mathematics, 1906 Association Drive, Reston, Virginia 22091.

National Microfilm Association, 8728 Colesville Road, Silver Spring, Maryland 20910.

National Science Teachers Association, 1201 Sixteenth Street, N.W., Washington, D.C. 20036.

National Tool, Die and Precision Machining Association, 9300 Livingston Road, Washington, D.C. 20022.

Pathescope Educational Films, 71 Weyman Avenue, New Rochelle, New York.

Paul Wallach Metric Aids, 2858 Carolina Avenue, Redwood City, California 94061.

Pauper Press, P.O. Box 303, Two Rivers, Wisconsin 54241.

Polymetric Services, 4600 Brewster Drive, Tarzana, California 91356.

Random House, Incorporated, Order Entry Department, Westminster, Maryland 21157.

RapiDesign, Incorporated, P.O. Box 6039, Burbank, California 91510.

Real-T-Facs, P.O. Drawer 449, Warwick, New York 10990.

Robie Sales, 2755 Woodshire Drive, Hollywood, California 90068.

Society for Visual Education, Incorporated, c/o Kenneth E. Clouse, 223 Quail Hollow Road, Felton, California 95018.

Spectrum Educational Suppliers, Limited, 9 Dohme Avenue, Toronto, Ontario, Canada.

Swani Publishing Company, P.O. Box 248, Roscoe, Illinois 61073.

METRIC JOURNALS AND NEWSLETTERS

Air Metri-Gram, American Institutes for Research, P.O. Box 1113, Palo Alto, California 94302.

American Metric Journal, AMJ Publishing Company, Drawer L, Tarzana, California 91356.

Metric Association Newsletter, Sugarloaf Star Route, Boulder, Colorado 80302.

Metric News, P.O. Box 248, Roscoe, Illinois 61073.

Metric Reporter, American National Metric Council, 1625 Massachusetts Avenue, N.W., Washington, D.C. 20036.

Metric System Guide Bulletin, J. J. Keller and Associates, Inc., Neenah, Wisconsin 54956.